Introduction to
VLSI Process Engineering

Introduction to VLSI Process Engineering

Edited by The Society of Chemical Engineers of Japan

SPRINGER-SCIENCE+BUSINESS MEDIA, B.V.

English language edition 1993

Originally published by Chapman & Hall in 1993

Softcover reprint of the hardcover 1st edition 1993

The English language edition has been translated from the original Japanese publication *Introduction to VLSI Processing for Chemical Engineers*, © Baifukan Co. Ltd (1989).

Typeset in 10/12 Times by Best-set Typesetter Ltd., Hong Kong

ISBN 978-94-010-4682-4

A catalogue record for this book is available from the British Library

Library of Congress Cataloging-in-Publication data

Introduction to VLSI process engineering / edited by the Society of Chemical Engineers of Japan; editor in chief, Yuji Naka. — English language ed.

p. cm.

Includes index.

ISBN 978-94-010-4682-4 ISBN 978-94-011-1562-9 (eBook)

DOI 10.1007/978-94-011-1562-9

1. Integrated circuits—Very large scale integration—Design and construction. 2. Manufacturing processes. I. Naka, Yuji, 1945– II. Kagaku Nōgaku Kyōkai (Japan)

TK7874.I5957 1993

621.39′5—dc20 92-45674

CIP

Printed on permanent acid-free text paper, manufactured in accordance with the proposed ANSI/NISO Z 39.48-199X and ANSI Z 39.48-1984

Contents

Editors and contributors

Editor-in-Chief

Yuji Naka
Tokyo Institute of Technology, Research Laboratory of Resources Utilization, 4259 Nagatsuta, Midori-ku, Yokohama 227, Japan

Editors

Katsuro Sugawara
Hitachi VLSI Engineering Co., 1-280 Higashi-Koigakubo, Kokubunji, Tokyo 185, Japan

Colin McGreavy
Department of Chemical Engineering, The University of Leeds, Leeds LS2 9JT, UK

Contributors

Tomonori Iwata
Tokyo Metropolitan Technical College, 1-10-40 Higashi-Ohi, Shinagawa-ku, Tokyo 140, Japan

Hideo Kameyama
Tokyo University of Agriculture and Technology, 2-24-16 Nakamachi, Koganei, Tokyo 184, Japan

Hiroshi Komiyama
University of Tokyo, 7-3-1 Hongo, Bunkyo-ku, Tokyo 113, Japan

Masakuni Matsuoka
Tokyo University of Agriculture and Technology, 2-24-16 Nakamachi, Koganei, Tokyo 184, Japan

Motosuke Miyoshi
Toshiba Co., 72 Horikawacho, Saiwai-ku, Kawasaki 210, Japan

Kunio Nagahama
Tokyo Metropolitan University, 1-1 Manami-Osawa, Hachioji, Tokyo 192 03, Japan

Masayoshi Ohno
Toshiba Co., 1-1-6 Uchisaiwai-cho, Chiyoda-ku, Tokyo 100, Japan

Tetsuo Onikura
NEC Co., 5-7-1 Shiba, Minato-ku, Tokyo 108-01, Japan

Hideki Tsuge
Keio University, 3-14-1 Hiyoshi, Kohoku-ku, Yokohama 223, Japan

Preface to the English language edition

The aim in producing the original version of this book was to provide a chemical engineering perspective of the technological problems in VLSI manufacturing operations, specifically in the context of Japanese industry. It was an initiative of the Society of Chemical Engineers, Japan, which brought together a group of engineers actively involved in developing the technology as well as being responsible for manufacturing operations. An emphasis on the manufacturing and operational problems makes this a distinctive contribution to the literature. The driving force is naturally directed towards meeting the needs of the Japanese domestic market, but the generic technologies are universal.

For the process engineer the subject matter introduces a technology which poses a new range of problems not normally encountered in the chemical industry. The treatment attempts to focus on relating the underlying physico-chemical principles to the development of manufacturing processes and the molecular level. It demonstrates how engineering problems can be tackled, and draws attention to the implications of working in a domain where microscopic detail is of crucial importance. This highlights a whole new class of difficulties which must be overcome as the industry moves forward to meet ever more demanding specifications. The examples and case studies are largely drawn from Japanese industry so that the estimates of marker size are naturally expressed in yen (¥). It is appropriate to retain this monetary unit for the domestic market, but where international markets are being discussed the US dollar is used; during the period of time covered by the figures the exchange rate was $1 = ¥130.

The Japanese edition was naturally based on references taken mainly from Japanese publications and they have been retained, since they represent the definitive sources of much of the content. They have been supplemented by some English language citations so that background knowledge may be acquired from sources to which readers of this edition should have easier access. In particular, it will be necessary to consult related texts on the underlying materials science. It is important to note that the coverage here is not intended to be as detailed as a handbook but rather the presentation of a concise account of the problems to be solved in developing a VLSI manufacturing facility. A particularly notable feature, however, is the attention drawn to the challenge of producing very high purity chemicals and the associated environmental constraints which must be met by this industrial sector.

VLSI manufacturing brings a new range of rapidly changing molecular processing technology problems to the attention of process engineers. It presents an exciting challenge demanding new approaches. The intention is that this monograph will provide a very useful introduction to the relevant process engineering aspects.

Preface

Every industry has in some way been affected by integrated circuits; not only are they used in computers but they now play an important part in cars, manufacturing machinery and scientific instruments. Greater varieties of integrated circuits are being put on to even smaller chips and the ever-increasing level of integration has meant that the demand has grown to meet the wider range of applications which have become possible. To achieve high levels of integration calls for very fine resolution of the basic elements in the circuit. In the case of the dynamic random access memory (DRAM), for example chips having 4 to 6 million memory cells covering an area no larger than the tip of a finger are now routinely produced. However, in order to be able to do this, the line-width memory cells must be less than 1 µm (10^{-6} m).

All of this has meant that the growth in the semiconductor industry during the last decade, for example, has been much greater than that of the heavy chemical industries in the 1960s. In Japan, growth halted in the period 1989–1990 because of the high value of the yen, trade friction with the US and general over-supply of semiconductors. Nevertheless there is still a strong demand for semiconductors from many industrial sectors and a high growth rate can be expected in the future.

Developments in manufacturing technologies in the semiconductor industry have made it possible to diversify the product range and so open up new areas of applications. It has now reached the stage where further growth will be dependent on improvements in micro-lithography and mass production techniques in order to make it possible to open up markets for highly functional special products.

In addressing these problems, it is useful to take note of the considerable expertise of the chemical industry in mass production technology. It too is shifting towards high added value products, including functional materials, and away from commodity chemicals and so placing an emphasis on small scale specialist processes to complement continuous large scale operations. There is, therefore, a commonality of interest in both VLSI and the increasingly important speciality chemical manufacturing.

In contrast, the basic technology in VLSI production is different from that of the petrochemical industry. In particular, quality specifications are very demanding in microelectronics because of the crucial role impurities play in determining the functionality of the integrated circuits. To meet the specifications it is necessary to adopt revolutionary new designs requiring novel operating procedures and support facilities.

Manufacturing products with specific functionality to the highest quality standards is extremely challenging, not least because the product specifications are not directly controlled. The final product may be used in a variety of applications. Consequently, it is necessary to be able to find an unambiguous relationship between operational variables and the desired specific characteristics of the product so that appropriate operational procedures can be devised which will ensure these can be met. The requirement is for systematic approach to designing products having a prescribed functionality which embraces all the manufacturing facilities and production procedures. To do this, there must be a coherent set of concepts relating research and development to design in order to be able to answer the following questions:

- Which manufacturing technologies are appropriate for highly functional products?
- What are the differences between this new technology and that currently employed?
- What are the principles which determine how to combine various basic process elements?

There is no easy answer to these questions and in any particular case it will depend on the desired characteristics of the products.

With specific reference to VLSI, a working group of the Society of Chemical Engineers, Japan looked at the technological issues. The results of this appraisal has now been compiled into this book. It was apparent that many chemical engineers have some knowledge of parts of semiconductor process technology but they have not been generally involved in the overall design of the process. In trying to bring everything together it is useful to start by considering the market, history and future of VLSI. This then sets the scene for presenting the design principles for devices, the overall VLSI process and the essential manufacturing technologies and inspection procedures. To complete the picture it is necessary to discuss the unresolved problems of the current production processes. This scenario provides the framework within which the chapters of this book have been organized.

The problems originating from the characteristics of silicon crystals, thin films and surfaces have been investigated in both materials science and chemistry. Until now there has been little consideration given to the manufacture of materials with the molecular and electronic functionality required by the electronics industry. In particular, VLSI processes make extensive use of micro-lithography, which rely on the control of mass and heat transfer to ensure that the desired product functionality is achieved. In chemical engineering terms this focuses attention on reaction engineering and transport processes. But it is also necessary to take account of the constraints imposed by working on a microscopic scale with the need to pay particular attention to product quality. The final chapter is devoted therefore to highlighting the ever-increasing range of problems in VLSI processing as the scale of integration grows, and showing the contribution that the expertise chemical engineers can make to providing a solution.

1

Coming age of VLSI

1.1 FUNCTIONALITY

Very Large Scale Integration (VLSI) is a term usually applied to solid state micro-electronic devices containing more than 64 Kbits of information on a single chip. Lower densities would be regarded as large scale integration (LSI). The actual size depends on the packing density and this in turn is governed by the line width of the circuit connections. This dimension is used as the criterion in what is termed the design rule, which for the case just quoted is 3 microns. To achieve the currently (1991) available densities of 2 million bits per square centimetre it is less than 1 micron and even higher densities are under development. The availability of such compact electronic devices offers the prospect of a wide range of new products which can exploit the powerful data processing capability in a convenient form for many new applications.

The transistor arose out of research concerned with improving the performance of telephone exchanges so that telephones would become more widely available. However, its potential use in applications other than simply replacing vacuum tubes (electronic valves) was recognized, especially because of the extremely high reliability and low electrical power consumption. Replacing transistors by an IC has made it possible to manufacture devices more easily so they can now be used to control domestic appliances electronically. These changes started in the 1970s and gave rise to the first generation of microcomputers typified by the Intel 4004. Some appliances and equipment which benefited from this were

- 1971: colour TV, pipe-organs and electronic clocks,
- 1974: 33 mm electronic cameras and microwave ovens,
- 1975: home VTRs, game machines, electronic dish washers, air conditioners and home sewing machines,
- 1976: autoselection stereo sound systems, personal computers, washing machines and 8 mm cine-camera,
- 1978: blood pressure meter.

During this time, LSI was mainly used for timing rather than digital computation because the limited capabilities of the four-bit processors were better suited to such tasks.

The move to VLSI has resulted in an enhancement of the capability of devices especially in the data processing functions. It has made it possible to use deductive reasoning to extend the range of tasks which can be undertaken. An analogy can be drawn with the human nervous system. Although it is still not possible to completely emulate the brain, VLSI devices are capable of controlling the senses. Of the five senses (sight, hearing, touch, taste and smell) the first three can be mimicked by semiconductor elements. Thus, sight can be replaced by photo-cells and light emitting diodes using a solid image sensor device such as the CCD (charge coupled device) which comprises arrays of up to four hundred thousand elements. However, even on this scale, the resolution tends to be poor. When compared with CRT displays, it is difficult to manufacture large screens having acceptable resolution, so such equipment has had restricted use. On the other hand, semiconductor devices have been used in microphones and receivers. In this case, quartz crystal detectors and magnetic induction are the primary information sources and VLSI is used only for amplification, processing and transmission of the signals. Speech synthesis using VLSI is possible and has considerable potential, for example in controlling cars and home robots.

Artificial touch is being intensively and extensively developed for temperature and pressure sensors. These devices have better sensitivity than those based on other principles and in fact have a wider range of application than would be appropriate for human beings. Temperature sensors are found in home appliances such as heaters as well as heated blankets, air conditioners and refrigerators and they can be combined with voice synthesizers to give warnings that normal temperature has been exceeded. Pressure sensors can be used as a basis for blood pressure meters. Sensors are also being developed to monitor moisture and fuel gases and can be expected to sense odours in due course.

The capacity of a memory device is expressed in terms of bits, which is an abbreviation for binary digit and is considered to be a basic unit of information. The term was first used by Shannon, while he was developing information theory. It makes use of plus and minus to represent yes and no. The human brain is regarded as handling huge amounts of information and is estimated to have the capacity of the order of 10^{16} bits. Clearly, VLSI would not be expected to be able to match this even with the further large increases in the number of bits in devices which can be expected in the future.

To gain some impression of the magnitude of numbers involved it is useful to take a simple problem to provide a frame of reference. For example, 2 Mbit would be needed to generate 280 000 image elements (pixels) on a black and white television screen (based on the 528 × 500 Japanese standard). This includes 8 bits, to represent 265 levels of shading between white and black (grey) for each of the pixels making up the screen. For a colour screen, the number needs to be increased by a factor of three to take account of the three primary colours i.e. a total of 6 Mbits. For high definition television pictures 30 Mbits will be necessary. Clearly, to provide this capability there is a need for even larger but more compact memories, otherwise the physical dimensions of the device limit its usefulness.

For use in computers, the memory must be even larger and give fast access, not only in the case of super-computers but for data-flow machines, neural network processors and high speed parallel computers.

1.2 USE IN CONSUMER PRODUCTS

Most homes have electronic devices which rely on ICs. In fact, the number of electrical motors per household has been used as a measure of the standard of living. In Japan this is of the order of 10–20. An alternative criterion could be based on the number of LSI devices. This has been estimated to range from tens to several hundreds, the latter being of the order of magnitude of LSI devices in a personal computer.

The digital world of the future can be expected to be based on intelligent devices which will bring with them a corresponding increase in the range of applications. For example, the electrical power consumption for refrigerators is 28.2% and for lamps 21.8%, colour televisions 12.5% and air conditioners 12%. To make energy savings, precise control of temperature of the refrigerator and brightness of a lamp is necessary which will require improved sensors that make use of intelligent control.

Electronic data processors comprising several VLSI devices are used in personal computers, word processors, tele-videophones as well as satellite broadcasting systems. The large, compact memory which is possible with VLSI opens up the possibility of new types of image processor. New types of television, such as IDTV, EDTV, high-vision, large TV wall-displays, non-ghost CATV, card-type and book-size videorecorders, stereo televisions with a printer for broadcast messages can be expected. In order to achieve all these high definition images or to have thinner displays or handle new media, vision systems using VLSI components need to be developed.

In the near future, recording of high definition TV will be possible using a digital VTR, which in turn means a demand for high definition video printers, stereo-TV and pocket VTR, video editors for home use and audio equipment supported with still and moving images. In addition, wireless-video systems can be expected along with digital audio tape

(DATs), digital audio-synthesizers, wireless (audio visual) AV systems and multi-AV-mixers.

There is considerable potential for developments in home-based information processing, including handwritten and speech-driven word processors, encyclopedias on CDs, workstations for domestic use based on 32-bit personal computers which have memory/editing systems for still and moving pictures together with high speed printers. In addition to these, there will be increased use of cordless telephones, television phones, memo-phones, digital HiFi and 'intelligent' domestic telephones. In all cases they can be expected to be more convenient to use than those available at present but will depend on being able to carry out more sophisticated information processing. Equipment such as microwave ovens with timers are already digital, but with more complete automation it will include security systems, automatic air conditioners, autovariable wall colouring, automatic room-lighting control and electronic notice boards. Of special interest would be robots for cleaning, looking after children and amusements (games). It will even be possible to compose music. The relatively low electrical power requirements will also make it possible to use such equipment almost anywhere.

1.3 IMPACT ON INDUSTRY

It is recognized that there is a growing demand for intelligent machines in industry and VLSI makes this feasible. There is every reason to believe that unmanned factories will be common and intelligent medical diagnosis machines will be developed. This will call for further developments in image processing.

The first generation of industrial robots first appeared in the 1970s. These carried out materials handling, painting and spot welding. The second generation in the early 1980s could sense and take decisions which gave a sense of 'seeing' and 'touching' which enabled them to be used for arc welding and simple assembly jobs. The late 1980s saw that robots could learn and so carry out general assembly operations, including movement between tasks.

The original motivation for such systems arose from the needs of industrial control systems which required repetition of simple tasks as for example in numerically controlled (NC) machines. These were based on IC or magnetic bubble memories. Nowadays, robots can handle objects very precisely using manipulators. The latest range of robots can measure the position of the work piece and align the job using seeing, touching, and hearing sensors, while third generation robots make use of developments in artificial intelligence in conjunction with improvement in sensors to carry out more sophisticated tasks.

There has been a change from factory automation (FA) through flexible manufacturing systems (FMS) and CAD/CAM to computer integrated manufacturing (CIM). FA is used in mass production lines found in the car industry. FMS is an integrated processing facility for transferring work and controlling flow of jobs to increase both productivity and achieve flexibility by making effective use of several NC machines. CIM unifies the whole management and control of labour, materials, process design and development, production, sales, financial affairs, and business within an infrastructure of computers. Each of these illustrates the vast and diverse range of applications which have become possible as a result of VLSI.

There is also an extensive range of applications in chemical and metallurgical industries especially in controlling production. In modern iron foundries, hundreds of sensors continuously monitor the plant and send huge amounts of data for processing. In research laboratories computers are used for molecular design in developing new materials and make particularly heavy demands on computer graphics.

In medical applications, expert system are used for diagnosis. The first example dates from the 1960s at Stanford University in the USA (MYCIN). Over the years, this program has been refined to take advantage of the experience of the doctors using it. In addition, there are applications in image processing (X-ray and computer tomography, CT) magnetic resonance imaging (MRI), nuclear magnetic resonance CT (NMRCT), ultrasonic diagnostic imager, thermography and digital radiography. The scanning tunnelling electron microscope (STEM) was developed in 1983 and resulted in the award of a Nobel prize in 1986. Its further development will inevitably depend on the continuing evolution of VLSI.

Car electronics are now largely dependent on microcomputer control and in particular the processing of data from a number of sensors. In the future anti-collision, anti-skid and car-navigation systems can be expected to become standard features, all of which rely on VLSI.

All of these examples show that the important part played by VLSI because it makes equipment compact and highly reliable. Developments in these areas will therefore depend on further improvements in semiconductor devices.

1.4 FUTURE TRENDS

The integrated circuit was invented by J. Kilby in 1958 and was subsequently improved by R. Noyce. Since then, two types of transistors (MOS and bipolar) have been developed although other types had been investigated. In the beginning, hybrid- and semiconductor thin film circuits were popular but subsequently were replaced by monolithic devices. Present technology makes it possible to make memory cell chips having ten million transistors per cm^2. The potential for future growth can be appreciated by noting that the 32-bit microcomputer requires only several hundred thousand which is two orders of magnitude lower. Personal computers use most of the transistors for memory cells. High speed calculations will need to replace the von Neumann architecture by new ones. One such development is the data-flow computer. This is more complicated and relies heavily on VLSI technology. Because the device is very reliable it can be expected to have a reasonable life span.

The main material used for VLSI is silicon and it is this which determines the processing technology. Switching speeds of the order of pico seconds are possible (80 ps). GaAs has also been investigated because of its faster speed, but it is difficult to fabricate devices using this so it has been restricted to opto-electronics. Intelligent systems based on VLSI technology can be expected to lead to computer graphic devices having large screens with precise stereo images, voice input word processors and computer terminals in gloves. Indeed, the range of possible applications is legion and limited only by the imagination.

2

Semiconductor market

Since its launch, semiconductor production has continued to grow rapidly, with the world-wide market of the order of $50 billion in 1990. This chapter examines those structural features of the industry which resulted in this high level of growth, market characteristics, the associated peripheral industries and their trends. It also considers the future prospects of the market.

2.1 FEATURES OF THE SEMICONDUCTOR INDUSTRY

Semiconductor components are essential components in a wide variety of industries and have been referred to as 'the oil of industry' in Japan. From the standpoint of manufacturers as distinct from users, semiconductor devices have the following characteristics: rapid technical innovation and high level of equipment investment.

Rapid technical innovation

Successive generations of memories, have demonstrated that the IC manufacturing industry is highly competitive and this has resulted in the development of new products, including highly integrated circuits, having enhanced functionality. Such rapid technical innovation demands high levels of investment in research and development (see Table 2.1)

In the Japanese semiconductor industry, investment increased from ¥21.5* billion in 1975 to ¥260 billion in 1985, i.e. a twelve-fold increase in 10 years. When compared with other major industries, the level of investment in 1985 ranked below that of the chemical (¥940 billion), motor car (¥800 billion) and mechanical engineering (¥380 billion) industries but was ahead of the steel industry (¥240 billion). The ratio of R&D investment to sales is very high and ranks alongside the medical

* The exchange rate against the US dollar is approximately $1 = ¥130.

Table 2.1 Ratios of R&D investment to sales in major industries

	1980 (%)	*1985 (%)*
All industries	1.5	2.3
Manufacturing industry	1.7	2.7
Semiconductors	12.6	14.1
Textiles	0.8	1.2
Chemicals	2.6	3.8
General chemicals	1.9	2.8
Pharmaceuticals	5.5	7.0
Steel	1.1	1.9
Machinery	1.9	2.7
Electric machinery	3.7	5.1
Electric machinery and appliances	3.4	4.8
Communication, electronic and electric measuring instruments	3.9	5.3
Motor cars	2.4	3.0
Precision machinery	3.0	4.5

Source: Statistic Bureau, Management & Coordination Agency *Science & Technological Research Report.*

equipment industry. The US has also seen a higher ratio than in other industries, suggesting that development of semiconductors requires a considerable R&D commitment. It is a trend which can be expected to continue in the future.

High level of equipment investment

Experience has shown that the cost of semiconductors follows a well-known learning curve which shows a decrease in costs of about 27% for a doubling in cumulative production. Thus, higher production can be expected to result in low production costs and a higher profit per unit encouraging large investment in equipment to facilitate mass production and sales. Unfortunately, this tends to cause a depression in the semiconductor market because of the oversupply arising from the excessive investment. On the other hand, there has also been increasing interest in ASICs (application specific ICs) for meeting particular applications and having special functionality. As a result, the practice of simply investing large amounts of capital in mass-production equipment has fallen out of favour. The level of equipment investment in major Japanese industry is examined in Table 2.2.

Table 2.2 Equipment investment by major industries (unit ¥100 million)

	1975 (%)	*1980 (%)*	*1984 (%)*	*1985 (%)*
Semiconductors	114	1 369	7 628	5 138
Electric power	15 432	33 090	31 734	31 927
Motor cars	3 366	10 998	9 907	13 797
Electronic equipment	1 905	6 496	11 474	10 276
Electric machinery	1 905	6 496	4 475	3 693
Chemicals	7 846	5 984	7 106	7 434
Steel	11 474	6 068	6 518	6 395
Industrial machinery	503	1 401	2 615	1 776
Total equipment investment	59 985	97 877	122 654	127 447

Note: Work basis (Source) Semiconductors: Based on data for 12 companies investigated by MITI (Ministry of International Trade and Industry).
Others: MITI "Equipment Investment Programs of Major Industries"

In 1984 and 1985, semiconductor industry equipment investment reached approximately ¥700 billion, almost on a par with the chemical and steel industries. Although a temporary decline followed because of the depressed semiconductor market in 1986, in the subsequent recovery of market conditions, equipment investment gathered momentum again and the ratio of equipment investment to sales has been maintained at a high level compared with other industries. When viewed by sector, the trend towards high investment in Japan is because of the high proportion of MOS memories in domestic products which are based on advanced micro-lithography. This may be regarded as reflecting the difference in commodity line-ups in the sectors. Semiconductor products having the above characteristics are considered in more detail below.

2.2 TYPES AND CLASSIFICATION OF SEMICONDUCTORS

Semiconductors are classified by individual components such as transistors, diodes, etc., and ICs (integrated circuits). An IC is defined as an electronic device which in itself performs a circuit function by integrating into a single unit various active and passive component parts which are not individually separable. ICs are further classified into hybrid and monolithic, the latter being the most commonly used type today (Fig. 2.1). Monolithic ICs incorporate active and passive parts on a semiconductor substrate and are classified as follows: operation and function; extent of integration; and general/specific use and standard/custom design.

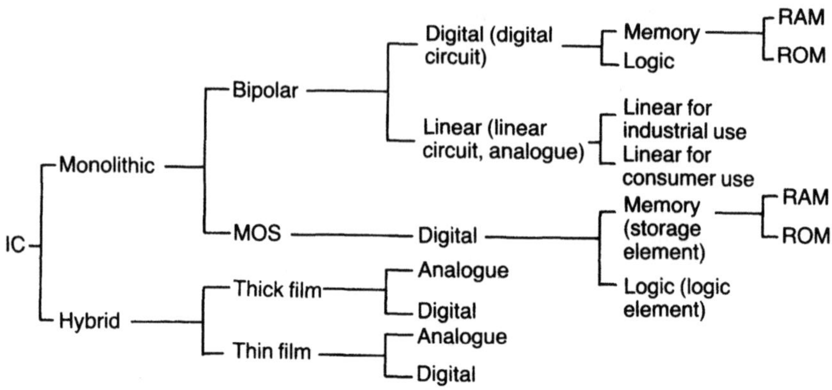

Figure 2.1 Classification of ICs by operation and function.

Operation and function

In terms of operation and function, monolitic ICs are classified as either MOS or bipolar. MOS ICs can be further classified as either memory (storage elements) or logic (processing elements). Bipolar ICs are classified as either digital or linear circuits. Bi-MOS technology has now reached the market, with MOS and bipolar ICs mixed on a substrate. The classification of ICs by operation and function is shown in Fig. 2.1 with the features listed in Table 2.3.

Extent of integration

ICs may also be classified according to the degree of integration, as shown in Table 2.4, with the boundary between LSI and VLSI being 10^5 elements, although in the case of logical ICs, 10^4 gates (several tens of thousands of elements) may be used as the boundary line.

General/specific use and standard/custom design

ICs can be classified according to whether they are designed and manufactured for a specific application (custom ICs) or whether they are for general use. Specific-use ICs are conventionally classified as full-custom ICs: general-use devices are called semi-custom. Those used in watches and games are examples of full-custom ICs, while gate arrays, mask ROMs etc. are examples of semi-custom. Specific-use standard ICs include those for particular applications, such TV sets and CD players. General-use standard devices include CMOS (complementary MOS) standard logic and dynamic memories. A further development has been

Table 2.3 Features and applications of major ICs

	Features	*Major applications*
Bipolar IC	Using bipolar transistor as an active element, resistor, diode and capacitor are incorporated in one silicon substrate. High-speed circuit operation. Large driving capacity. High analogue process accuracy. Medium to large power dissipation.	Computers (25%) VCRs (17%) Communication equipment (11%) Colour TVs (7.4%) World processors, personal computers (2.5%) Acoustic equipment IC testers
MOS IC	Composed mainly of MOS FETs as active elements. High degree of integration because of simple construction. Easy, low-cost manufacturing process. Less power dissipation. Medium operating speed. Small output current. Not suited to analogue function.	Computers (15%) Word processors, personal computers (11%) Communication equipment (9.3%) VCRs (5.6%) Colour TVs (2.3%) Acoustic equipment, NC equipment Industrial robots, watches, games equipment
Hybrid IC	Combination of single parts such as diode, transistor, capacitor with a film circuit. Combination of various elements having optimum electrical characteristics is possible. High-speed switching is possible. High-tension and large output are possible. Can provide digital and analogue functions. Economical when production quantity is small. Small degree of integration.	Acoustic equipment Computers IC testers Medical equipment Motor cars

Table 2.4 Classification by degree of integration

Name of IC	*No. of elements/chip*
SSI (small scale integrated circuit)	~100
MSI (medium scale integrated circuit)	100~1 000
LSI (large scale integrated circuit)	1 000~100 000
VLSI (very large scale integrated circuit)	100 000~

the concept of the ASICs (application specific ICs). It is not easy to give a definition of an ASIC, but a classification currently used is shown in Fig. 2.2.

Increased integration densities have reduced the general applicability of ICs. When a complete system is put on a silicon chip, it becomes a product for a specific application. From a functional viewpoint, composite microprocessors and memories have made it even more difficult to distinguish between ICs by function.

2.3 EXPANDING SEMICONDUCTOR MARKET

From the first, the semiconductor industry has expanded rapidly and according to a study by WSTS (World Semiconductor Trade Statistics), the world semiconductor market was about $50 billion in 1989, which represents an annual rate of almost 20% over the past 15 years. Of the $50 billion market, North America (including Canada) has 30%, Japan 39%, Europe 18% and Southeast Asia and other areas 12%, as shown in Fig. 2.3. Japan is the largest consumer, overtaking the US in 1986 despite the rapid changes in foreign exchange rates. However, this does not include several large scale semiconductor producers in the US and some semiconductor products which are consumed in-house and are estimated to be valued at more than $5 billion. If this is taken into consideration, the US is still the largest consumer.

Recently, the demand for semiconductors in the newly industrialized economies (NIEs), centred in Southeast Asia, has rapidly increased. This rapid growth in demand is indicative of the changing industrial structure in this area which fills gaps in the manufacturing industries of Japan, US and Europe.

Although the semiconductor market has undergone rapid growth in a short period, this expansion has not been continuous. Several depressions have been experienced due to external conditions or factors peculiar to

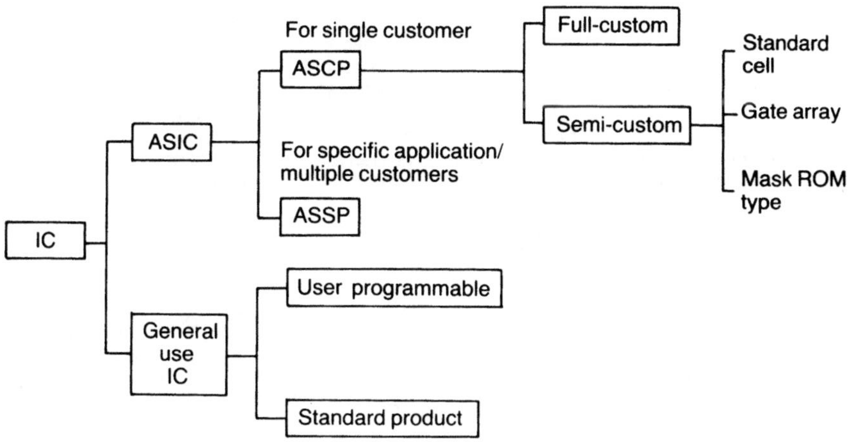

Figure 2.2 System diagram of ASIC.

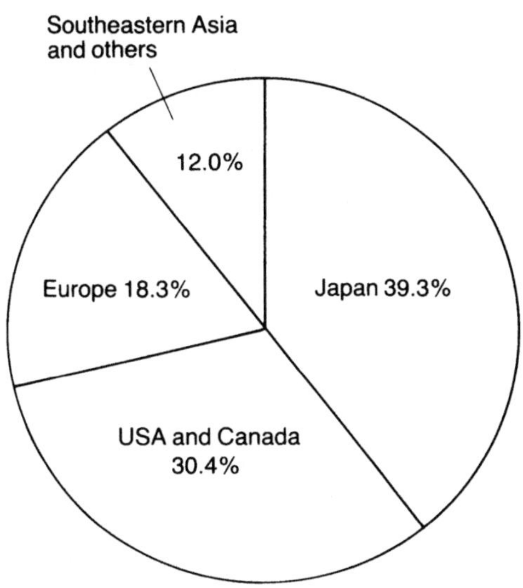

Figure 2.3 Breakdown of world semiconductor demand (1989, Source; WSTS).

the industry. The market boom–bust cycle, which tends to be repeated nearly every three or four years is generally referred to as the silicon cycle and is shown in Fig. 2.4. Business fluctuations, production of electronic equipment, semiconductor equipment life cycles and the like are considered as factors which give rise to this.

It is instructive to look at the semiconductor market in Japan. The level

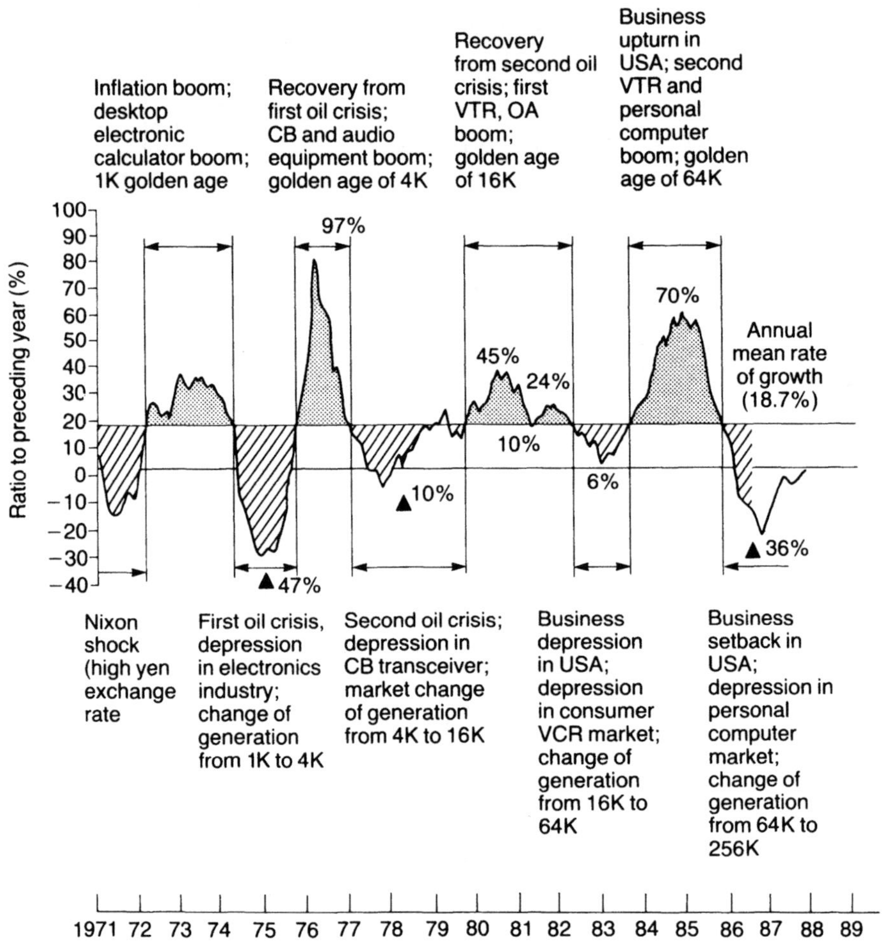

Figure 2.4 Silicon cycle.

of domestic demand in 1989 was about ¥2.0 trillion, while domestic production, exports and imports were ¥2.5 trillion, ¥680 billion and ¥170 billion, respectively, with exports exceeding imports by ¥510 billion. Historically, it is only relatively recently that exports have exceeded imports in Japan, with an adverse trade balance period up to 1978. The increase in exports has been possible because of factors such as the excellent quality of Japanese semiconductors, the large supply capacity and the demand for high technology products.

Figure 2.5 shows a comparison, by area, of the application-classified semiconductor demand structure. Japan has two major semiconductor application areas based on VCR/audio equipment and computer related

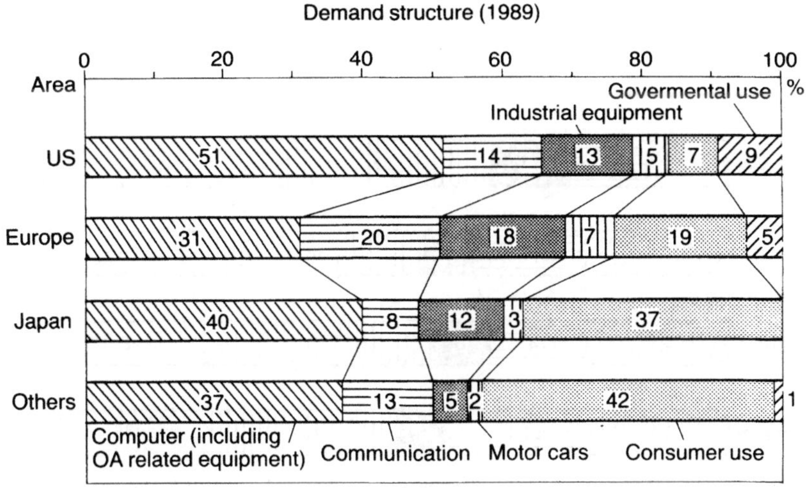

Figure 2.5 Semiconductor demand by area and application. (Source: Dataquest.)

products. In the US the use of semiconductor applications in computers and government (defence) is considerably higher. The demand structure in East Asia is almost the same as that in Japan. The difference between the two is in the ratio of applications in industry and communications. In Europe, the demand for semiconductors for consumer and industrial use, communications and computers is well balanced. In the circumstances. the semiconductor product structure differs slightly depending on the area (Fig. 2.6). In the US, the ratio of memory and bipolar digital ICs used for computer related devices is high. However, in the case of linear ICs used for consumer devices it is low. On the other hand, in East Asia the ratio of memories to linear ICs for consumer use is the highest, showing a demand structure which reflects the industrial balance in the area.

2.4 TRENDS IN ASSOCIATED INDUSTRIES

2.4.1 Semiconductor manufacturing equipment

The semiconductor industry is often referred to as the equipment or sometimes the facilities industry. This is because semiconductor manufacture involves various processes using a variety of manufacturing equipment with the result that progress in semiconductors is independent of progress in this sector. Figure 2.7 gives an indication of the world semiconductor manufacturing equipment market. In the early days of the industry, reliance on manufacturing equipment from the US was naturally high. After about 1980, however, rapid progress was seen in Japanese-

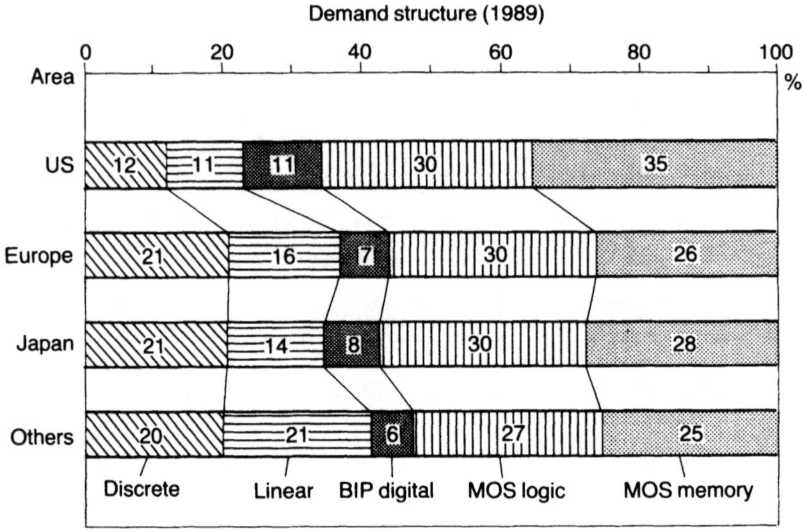

Figure 2.6 Semiconductor demand by area and product. (Source: WSTS.)

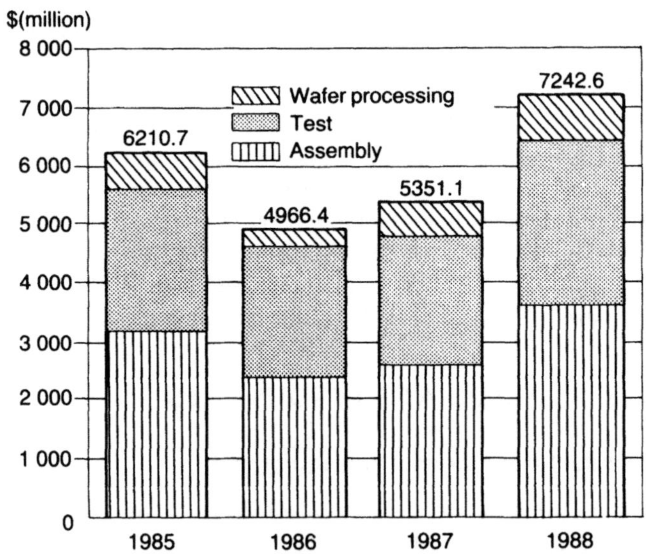

Figure 2.7 Worldwide semiconductor manufacturing equipment market (Source: VLSI Search Inc.).

made manufacturing equipment and Japan is now a world leader in many areas related to assembly, testing and wafer manufacture. Japanese equipment accounts for about 60% of the total at present, with the likelihood of increased domestic production in the future. The production of semiconductor manufacturing equipment in Japan was worth ¥280 billion in 1986, with half of this figure accounted for by wafer processing equipment. The world-wide market was $5 billion in 1986 and estimated to be $7.2 billion in 1988. This is equivalent to about 15% of the total production of semiconductors. The manufacturers are actively pursuing the development of advanced products and maintain a constant demand for manufacturing equipment requiring increasingly high levels of technology.

2.4.2 Semiconductor materials

The semiconductor materials industry had made advances in technology along with the growth of the industry, providing indirect support for the development of the business. Taking the silicon wafer as an example, the diameter of ingots has increased at a rate of 1 cm per year. This has resulted in an increased number of chips per work unit and to greater production efficiency. To meet the demand for increased size, there has been a requirement for materials with mechanical specifications having resistance to wear, warp and so on, as well as improved properties such as uniform electrical resistance and oxygen content. Silicon wafers are now generally 6 inches (15 cm) in diameter at present (1990), but this is expected to be increased to 8 inches (20 cm) and eventually to 10 inches (25 cm) as a result of R&D by the manufacturers.

The materials market also expanded along with the increase in semiconductor production, and is said to be worth about $650 billion in Japan, and $6.3 billion worldwide with wafers, photomasks and ceramic packages accounting for a high proportion of the total. See Fig. 2.8. Composite semiconductor wafers are now attracting attention as materials for the future. World-wide demand in 1986 was put at $135 million, consisting of $66 million for GaAs, $52 million for GaP and $10 million for InP. It is expected that GaAs devices will be used for high-speed computers in the future. There is also the problem of increasing the purity of doping materials with the further development of VLSI of devices, with a shift in grade from 5N (99.999%) to 6N (99.9999%). The estimated consumption of major dopants is shown in Table 2.5.

2.5 FUTURE PROSPECTS

It is useful to consider the market for semiconductor devices in the future. The demand is determined by the products which use them and

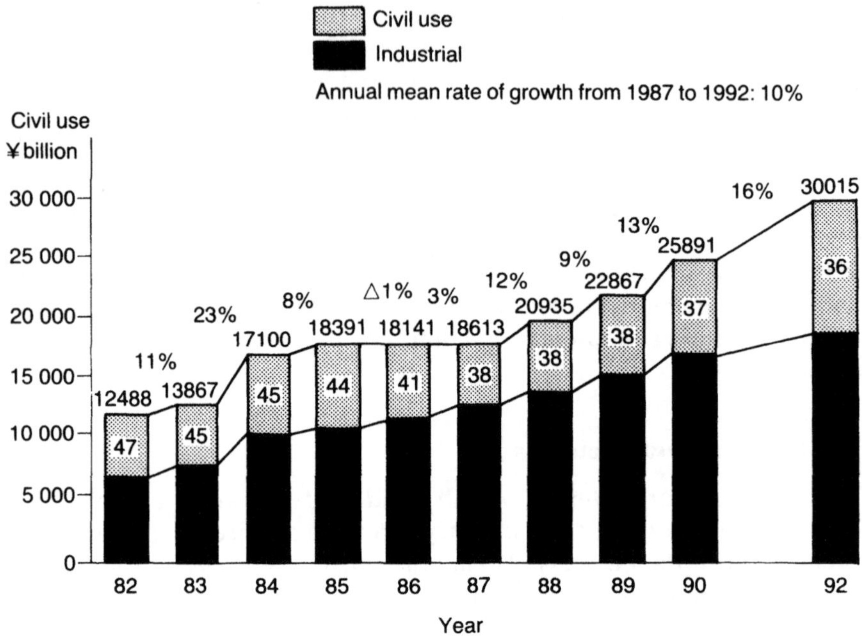

Figure 2.8 Actual and estimated amount of equipment manufacture (Source: Actual: EIAJ).

Table 2.5 Typical fraction of semiconductors in equipment (%)

	1982	*1987*	*1992*
IC	5.4	7.6	8.6
Individual semiconductor	2.3	2.3	1.9
Total	7.7	9.9	10.5

the size of the market for such products. Although overseas production is increasing steadily to cope with changes in exchange rates and to avoid trade friction, an average growth of about 10% per year is expected in the mid-term for equipment manufactured in Japan. This can be attributed to the increasing demand for digitally-based consumer products, developments in home automation, increased computer speeds, and use of AI (artificial intelligence). As a result, the value of the domestic equipment output is expected to reach ¥30 trillion in 1992. The ratio of the semiconductors used to equipment produced is called the semiconductor component per equipment ratio (I/O factor). This factor shows a gradual increase with the progress in the development of electronic, multifunc-

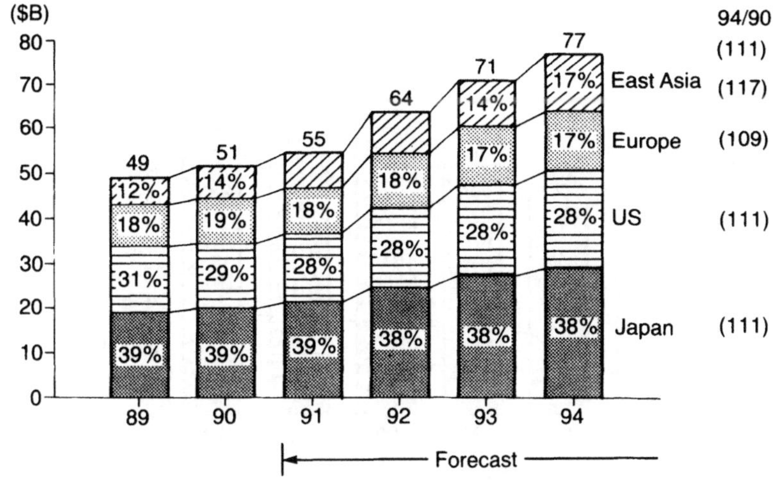

Figure 2.9 Worldwide semiconductor market trend (by area). (Source: WSTS.)

tional and high-performance equipment. The expectation is that this trend will continue in the future, with the I/O factor increasing to 10.5% in 1992. When the future size of the semiconductor market is calculated based on the above factors for expansion of equipment production and an increase of the I/O factor, a domestic demand of ¥3.2 trillion is expected in 1992. This figure is 1.7 times that for 1987, and represents an annual mean rate of growth of 11%. The world-wide semiconductor market shows a slightly higher rate of expansion than that of Japan partly because of the high growth rate of the Asian NIES and EC. According to the latest WSTS forecast, the world semiconductor market will amount to $75 billion in 1993 with Asian NIES and EC countries increasing their share at the expense of Japan and the US (Fig. 2.9).

WSTS estimates that memories, microcomputers and MOS logic ICs will increase their share, classified by product (Fig. 2.10). Though this is a high growth rate when compared with other industries, the semiconductor industry can be said to be gradually shifting to a period of steady growth, when compared with the 20% annual growth rate experienced in the past.

It is interesting to examine the future prospects in terms of the qualitative changes. The economy of the advanced nations has entered an age of maturity after passing through a period of fast growth and in general consumers are looking for a more comfortable lifestyle against a background of a stable economy. Consumers are looking more and more for products which reflect individuality rather than cheap, uniform, mass-produced items and this is resulting in major changes in the design of electronic equipment. In improving the features offered by earlier

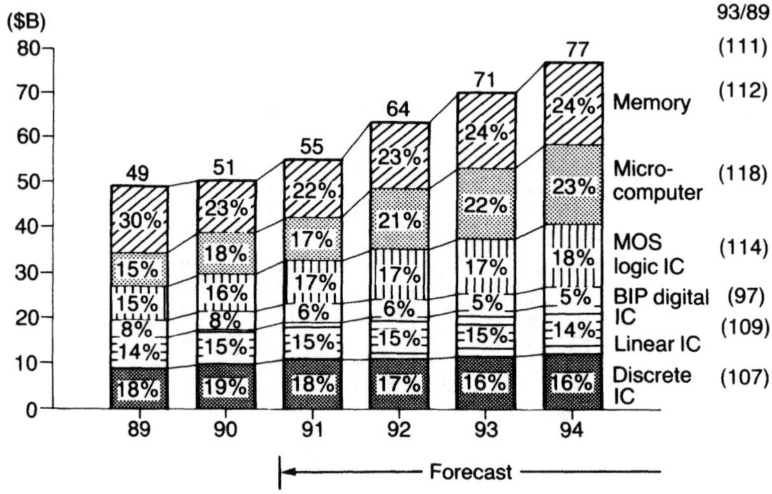

Figure 2.10 Worldwide semiconductor market trend (by product). (Source: WSTS.)

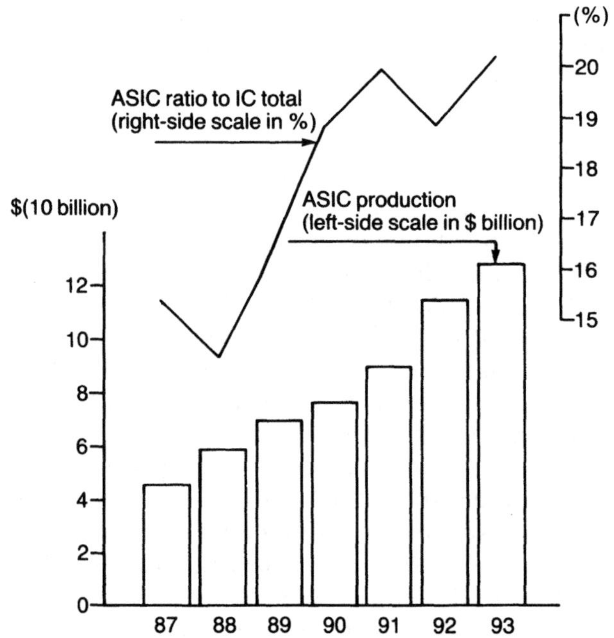

Figure 2.11 World ASIC production and fraction of total (estimate).

products, it will be increasingly important to pursue new, special features and develop products which meet a wide variety of needs are a matter of course. Further, product life cycles have been reduced, and the benefits of scale resulting from long production runs are no longer easy to attain; the trend toward diversified small-quantity production is clear. Against this background of change, the expected direction for future development of various semiconductors is seen as the ASIC (application specific IC). Through the introduction of the ASIC, users can expect differentiation from other manufacturers' products and the ability to handle diversified small-quantity production, together with such benefits as improved performance, quality and manufacture of small, lightweight products (Fig. 2.11). Advances in ASIC technology also make it possible to implement a large scale system on a single chip by merging the systems engineering of the user with the device technology of semiconductor maker at a high level, through engineering support such as CAD (computer aided design). This trend towards the use of ASICs is one answer to the increasing diversity of social need and will demand a transformation of the essential structure of the semiconductor industry from its past orientation towards mass production.

3

Evolution of semiconductors

Semiconductor devices have now entered the VLSI age and have found a wide range of applications in fields quite different from those of traditional electronics. This has meant that production has had to increase continuously to match consumption. The increase in applications has been due to the dramatic improvement in integration density, performance, reliability and cost reduction. The production of devices having a given integration density exhibits a so-called silicon cycle of approximately three or four years as a result of the introduction of new higher density products with better performance at reduced costs. Semiconductors show every sign they will continue to develop as a typical high-tech industry. In this chapter the history and perspective of semiconductor technology is explained.

3.1 HISTORY OF SEMICONDUCTOR TECHNOLOGY

There have been critical developments in materials and the closely related process technologies which have led to the present situation where the semiconductor industry is dominated by VLSI devices based on silicon crystals. It is helpful to look at the way in which these developments have taken place with respect to semi-conductor materials, processing and device fabrication.

Semiconductor devices are produced by fabricating multiple electronic functions on semiconductor materials such as Si or GaAs. This involves mainly physico-chemical operations, but it is also important to take account of the stability and reactivity of the materials. Figure 3.1 indicates the relation between the properties, processing and devices. The history of semiconductors is outlined in Fig. 3.2 and their main properties are listed in Table 3.1.

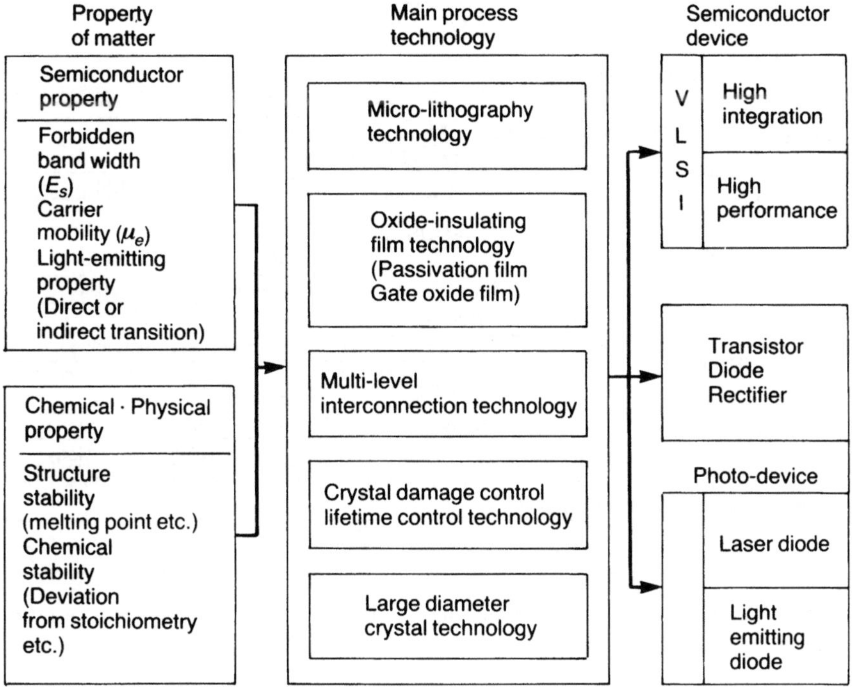

Figure 3.1 Relation between property of semiconductor material and semiconductor device.

3.1.1 Materials

In the latter half of the 19th century, Cu_2O and Se rectifiers were developed, but because of poor reproducibility and quality control they were eventually replaced by Ge, Si and Group III–V compounds.

Germanium

As can be seen in Fig. 3.2, point-contact transistors [1] were invented in 1948 at the Bell Telephone Laboratories followed by alloy-type transistors in 1949. The p-n junctions in the later Ge transistors had good reproducibility. In practice, this was achieved by the development of the alloy-junction method in 1950 using In, Ga or other dopants and by the zone melting method in 1952, which gave high quality Ge single crystals.

Silicon

Since Ge possesses a forbidden band width of 0.66 eV, there is a loss of semiconductor characteristics at relatively low temperatures, so the

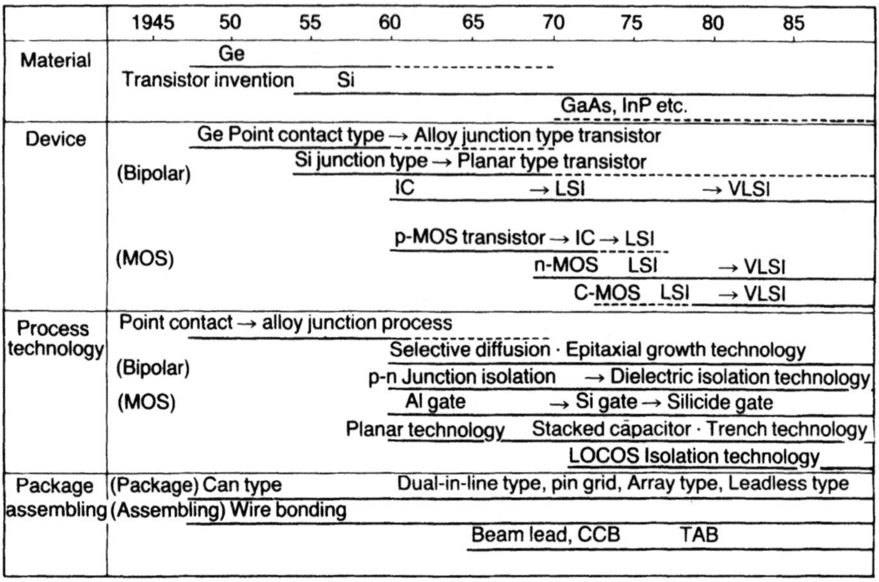

Figure 3.2 History of semiconductor: CCB = controlled collapse bonding; TAB = tape automated bonding.

maximum temperature of operation is limited to about 85°C. The semiconductor characteristics depends on the formation of p-n junctions by the addition of small amounts of dopants (impurities). Silicon was selected because it belongs to the same group as Ge and has a wider forbidden band width of 1.11 eV, which allows transistor action up to approximately 175°C. It is a good semiconductor material, not only because of the wider forbidden width, but also because it has been possible to develop the necessary processing operations for fabrication. In particular this includes the following.

1. *Selective diffusion* (1957): patterns of the SiO_2 films formed on the silicon substrate can be used as windows for selective impurity doping.
2. *Planar technology* (1960): thermally grown SiO_2 films on a silicon substrate are now widely used. The interfacial properties at the Si–SiO_2 interface ensures high performance and high reliability of the transistors.
3. *Growth of dislocation-free crystals* (1960): this has enabled the growth of large diameter high quality single crystals.

These could develop because the melting point of silicon is high and oxidation occurs at elevated temperatures so the resultant SiO_2 films are very stable. The development of micro-machining has also played an important part in fabricating transistors for, ICs, LSI and VLSI.

Table 3.1 Material properties of Ge, Si and main III–V compounds

Constant \ Material		*Si*	*Ge*	*GaAs*	*InP*	$In_{0.71}Ga_{0.24}As_{0.55}P_{0.55}$	$In_{0.65}Ga_{0.35}As_{0.79}P_{0.21}$
Crystal structure		Diamond type	Diamond type	Zinc blende type	Zinc blende type	Zinc blende type	Zinc blende type
Density [g/cm^3]		2.3289	5.323	5.3161	4.789	–	–
Forbidden band width (300 K) [eV]		1.11	0.66	1.44	1.25	0.95	0.80
Mobility (300 K) [cm^2/V s]	Electron (μ_e)	1800	2800	8.650	5000	–	–
	Positive hole (μ_h)	400	1800	450	160	–	–
Melting point [°C]		1410	937 ·	1238	1058	–	–
Application		VLSI etc.	Long wave photo-detector etc.	Micro-wave device etc.	Light-emitting diode etc.		

Many improvements in materials and process technologies were required in developing micro-lithography which was needed to achieve higher integration densities and performance. Miniaturization of devices necessarily requires thinner films and junctions in accordance with the so-called scaling rules (see Section 4.1.4). New materials for both metallization and insulation are required to construct the devices. Three dimensional structures of the silicon wafers are essential in realizing higher density and speed. The R&D underpinning such new technology needs to focus on the following.

1. *Development of new materials*: building VLSI devices relies on materials for metallization and insulation. The former is necessary for the connections and the latter for capacitors which are used in memory cells. Commonly, aluminium alloys have been used for metallization; however Al-layered structures, tungsten, other refractory metals and copper are now being considered because of the move to smaller dimensions and thinner junctions. They are useful for micro-fabrication because they permit good ohmic contact with thin junction Si substrates as well as preventing the electromigration with high currents through the fine grain metallization. Al easily diffuses into Si at high temperatures and causes current leakage in thin junctions. Consequently there are a significant number of problems to be solved before such materials can be used for large scale production.

   Although stable SiO_2 and Si_3N_4 films have been used as insulating materials for capacitors in memory cells, new materials having higher dielectric constant values, such as Ta_2O_5, have been considered for miniaturized large capacity VLSI RAM. This is because VLSI memory requires very small capacitor dimensions, which is possible with Ta_2O_5 films, since the dielectric constant is six times larger than that for SiO_2 films. Figure 3.3 shows the relation between thickness of the insulating film and cell area when the cell capacitance is kept constant. In the case of 16Mbit RAM, the required thickness of the SiO_2 film would be about 3 nm while that with Ta_2O_5 would be 18 nm. This considerably reduces the electric field intensity across the film.
2. *Thinner films and junctions*: to miniaturize the device, smaller transistors in the VLSI must be created. The performance of memory cells can be improved with thinner gate oxide films for the MOS transistors and thinner junctions for the source and sink. The trend is towards very thin gate oxide films of less than 10 nm. This requires high reliability, which relies on improvements in film formation, a higher selectivity ratio during the formation of the gate electrodes and less damage of the electrical characteristics of the devices. To achieve thinner junctions (of the order of 0.1 micron) ion implantation using a large current, low acceleration voltage and low temperature annealing

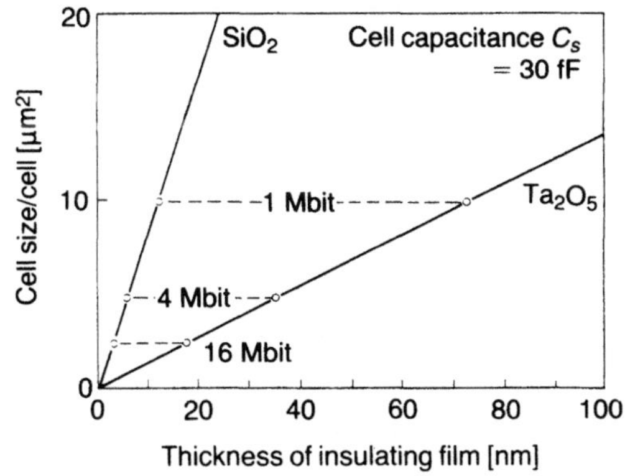

Figure 3.3 Relation between thickness of insulating film of various dielectric constant and memory cell size.

is necessary. At present, annealing at temperatures higher than 900°C is essential to activate the implanted impurities. Generally there is a need to improve the annealing operation.

3. *Three-dimensional structures*: present VLSI devices are, in principle, two-dimensional devices since patterns of the device are written and components formed on the silicon wafers. Micro-fabrication and high levels of integration have therefore essentially been accomplished by reducing the device areas.

   Future advances are expected to be directed towards forming devices having multifunction, high speed and high integration density fabricated in a three-dimensional structure. Various choices are possible. For example, it is possible to form grooves or to deposit multi-layer films on two-dimensional devices. Alternatively, sets of components in multi-layers can be formed. Some attempts have been made to introduce three-dimensional structures in two-dimensional VLSI devices smaller than 16 Mbit RAM. Typical examples are the isolation of components such as capacitors using stacked structures, deep grooves or trenches in Si wafers, as shown in Fig. 3.4. Alternatively, multi-level metallization can be used to give greater integration and higher speed. It is necessary to improve planarization technologies to avoid reduced reliability caused by the uneven surfaces on Si wafers which arise with fine multi-level metallization. The technologies includes a refill process of the metal in the very small through-holes and planarization of the insulating layer. One way of doing this is to form a planarized insulating layer on an unevenly metallized surface, using a biased ECR method. This is a new technology capable of forming sub-micron metallization films having an aspect ratio greater than 1.0. It

Memory cell	Manufacturing	Cell structure	Problem
Planar capacitor	Relatively easy	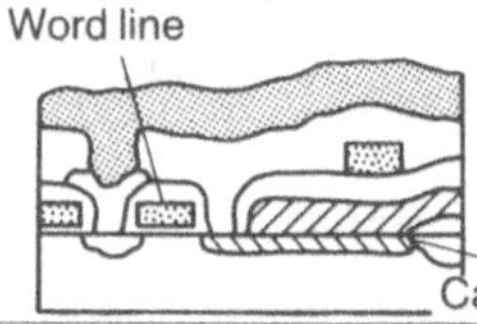	Very thin SiO_2 film (< 6nm)
Stacked capacitor	↓ Fairly difficult	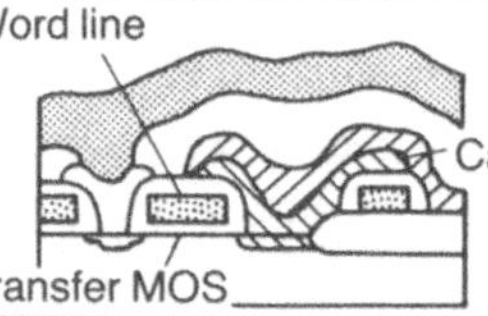	Fabrication of high aspect ratio construction with very thin insulating film
Trench capacitor	↓ Relatively difficult	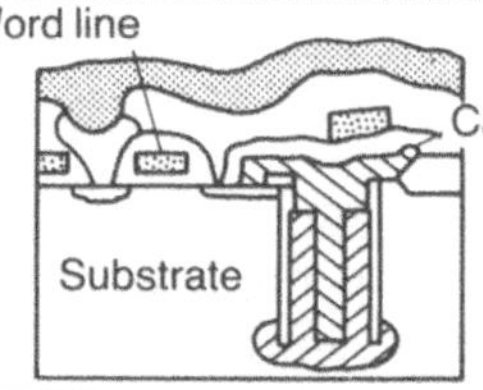	Trench fabrication (sub μm) Isolating in trench Stacked capacitor in trench

Figure 3.4 Memory cell designed for megabit DRAM.

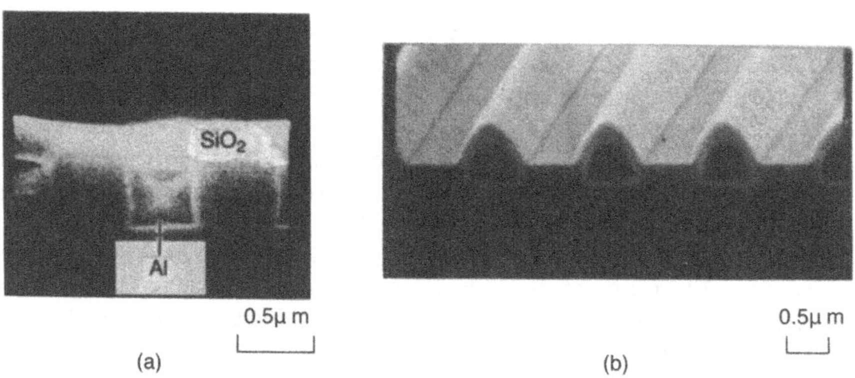

Figure 3.5 Planarization technology of intermetal insulating film used for multi-level metallization. Step coverage formed by: (a) biased ECR SiO_2 film; (b) conventional CVD film.

involves a combination of deposition, where the particles impinge on to the vertical Si wafers and conventional sputtering. Figure 3.5(a) is a cross-sectional photomicrograph showing a planarized biased ECR SiO_2 film deposited on an Al metallization having a line a width of 0.5 micron. The improvement is remarkable compared with films prepared using the standard procedures (see Fig. 3.5(b)).

Essentially, a new technology for growing a single crystal layers of silicon on the insulating film (called silicon on insulator (SOI)) is necessary to build up sets of components in multi-layer structures. It is difficult to obtain layers of good crystals at present and various investigations are now under way to increase the density and minimize the metallization length. Three-dimensional devices such as photoelectric conversion sensors and image processing circuits are of particular interest, since they are multifunctional device components. To reach this stage will require further developments, including significant innovations in the processing, before truly three-dimensional devices can be formed.

Semiconductor materials of group III–V compounds

Silicon has been widely used in the production of LSI and VLSI. However, group III–V compounds are potential semiconductor materials because:

- the transition of electrons from the conduction to the full band is direct and results in high efficiency light emission;
- different compositions of the materials give rise to changes in the width of the forbidden bands, resulting in changes in the wave length of the emitted light; and
- the compounds have high electron mobilities.

Because of these attractive properties these compounds have been the focus of intensive research activity. They are quite different from those of Si and Ge, so the emphasis has been on the development of opto-related and super high speed devices, of which light emitting diodes, laser diodes for optical telecommunication, compact discs and video disc players are typical examples. Figure 3.6 shows a section through the multi-level structure of a buried heterojunction [2] near the active layer. It is a 1.55 micron band laser diode used for optical telecommunication. An InGaAsP layer is formed on the p-type InP substrate during epitaxial growth. The multi-level structure is formed by the repeated etching and epitaxial growth. When a current flows under forward bias in the laser diode, a laser beam is emitted normal to the plane of the section in the In-GaAsP active area.

It is useful to make a comparison between the three semiconductor

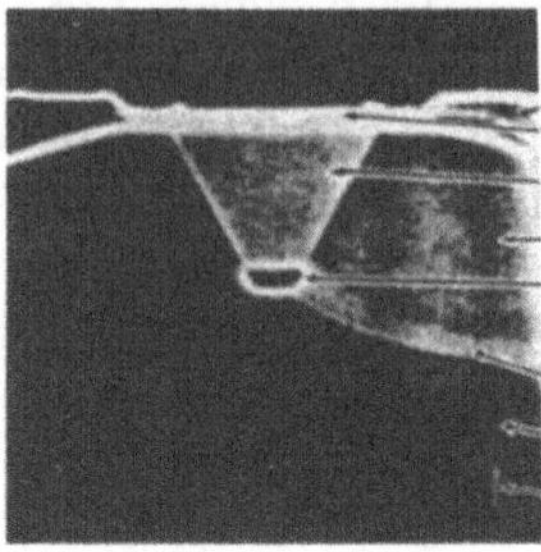

Figure 3.6 Cross-section of 1.55 μm band laser diode [2].

materials, Ge, Si and the group III–V compounds. Si has replaced Ge since it has better basic semiconductor properties such as the forbidden band width and is structurally stable. The latter arises from the fact that the melting point of Ge is 937°C, which is lower than that of Si (1410°C). A lower temperature imposes limitations on the fabrication technology. The group III–V compounds have better semiconductor properties than Si, including the forbidden band width, carrier mobility and light emitting characteristics. However, there is only limited scope for developing a manufacturing technology which is capable of making VLSI using such compounds on a mass production scale and so they offer no threat to the position of Si. Consequently, Si is still used in large quantities as the major source of VLSI devices where micro-lithography and planar technology can be easily applied. The present situation means that the group III–V compounds can only be used as optical or super high speed devices, which is approximately 1% of semiconductor crystal production.

3.1.2 Process operations

The basic device in VLSI is the metal oxide semiconductor (MOS). Micro-lithography and high density packaging are the most important manufacturing technologies which make it possible to achieve high levels of integration, performance, reliability and economy, as is illustrated in Fig. 3.7.

The methods which have made it possible to realize higher levels of integration are as follows [3]:

- development of micro-lithography to reduce the size of patterns and of components such as the memory cell;
- decreasing the defects in single crystal will make the effective area of the Si chip larger, though there is a trade-off between the fineness of the pattern and the yield; and

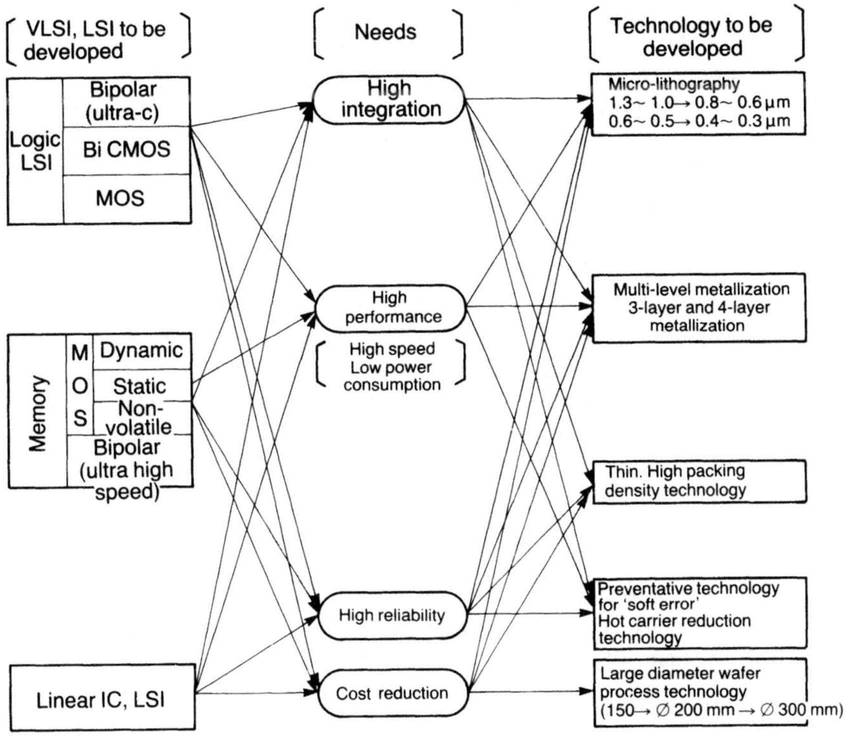

Figure 3.7 Device and technology to be developed.

- improvement in the performance of devices and circuits which increase the degree of integration.

As shown in Fig. 3.8, integration has increased by a factor of about 4 every three years. This tendency is expected to continue for the next few years, though at a slightly lower level. In the light of this, it is useful to review the existing problems of silicon VLSI manufacturing. Table 3.2 shows the expected future development of VLSI and it is clear that the micro-lithography will be important.

As will be seen later, micro-lithography consist of various important processes which need to be balanced with each other. Planar technology is an example which uses an SiO_2 film as insulator to give chemical stability. Other approaches have been developed to increase the level of integration. SEM pictures of a MOS DRAM going from 64Kbit to 4Mbit are shown in Fig. 3.9. This corresponds to the feature size decreasing from 3 to 0.8 μm. The stages involved in micro-lithography are as follows.

1. *Mask making*: This operation uses an electron beam exposure method instead of the conventional optical method. The change has improved the accuracy and reduced the defect density of the masks.
2. *Aligner technology*: This covers the alignment of components and interconnection patterns on a Si wafer during micro-lithography. It has brought about a change in the reduction ratio of 1-to-1 to 5-to-1 projection. Employing so-called self-aligned technology it is possible to partially use an existing film instead of a conventional resist.
3. *Resist material*: positive resists has been introduced. The parts exposed to radiation are soluble in the developing solution but the negative resist is insoluble after exposure. This makes it possible to achieve a more precise patterning than with negative resists.
4. *Etching*: wet etching in solution has been replaced by plasma etching. Plasma etching makes possible anisotropic reaction which results in improved accuracy.
5. *Thin film formation*: materials have been developed for insulators in multi-level metallization and passivation films. Both procedures give good quality films.
6. *Impurity diffusion*: The effectiveness of concentration control of the impurity dopant has been improved by replacing the conventional thermal diffusion technology by ion implantation and annealing.

Two other important processes need to be employed in order to achieve higher levels of integration: micro-lithography and dry etching.

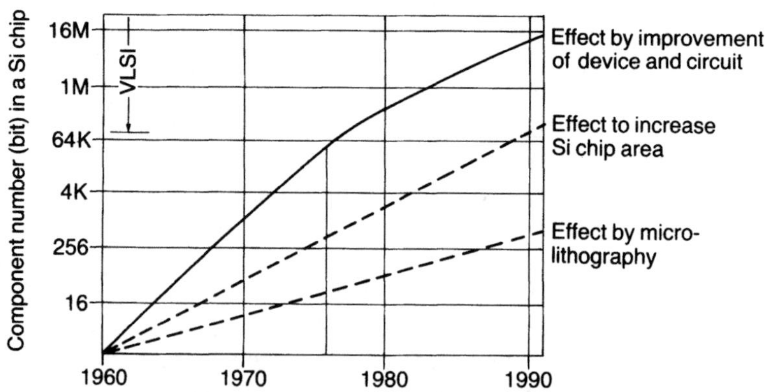

Figure 3.8 Trend of device high integration and its key factors [3].

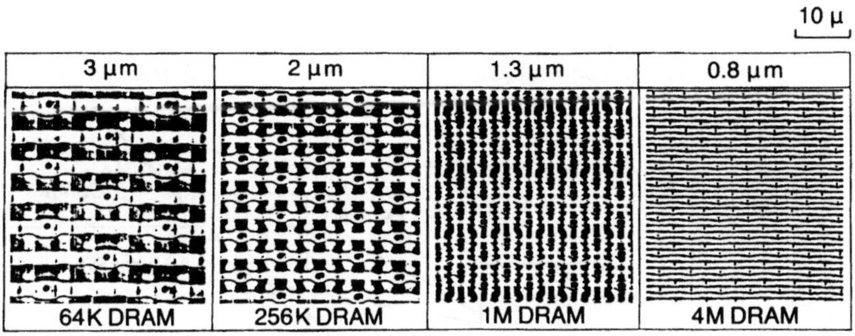

Figure 3.9 Patterns of five generation MOS memories observed by SEM.

Table 3.2 Development of VLSI process technology

Target	*Method*	*Element technology*
	Micro-fabrication	Lithography – optical, electron beam, X-ray, SOR Etching – high accuracy, high speed, one wafer etcher
High integration,	Development of new material	Conductive material – refractory metal silicide Insulating material – material of high dielectric constant
High speed, High performance	Thin film formation – thin film, shallow junction	Low temperature process – plasma, photo-chemical reaction Low damage process – low energy process, photo-chemical reaction
	Three-dimensional structure – multi-layer structure	Trench structure – isolation, capacitor, transistor Multi-layer metallization – planarization, selective deposition, taper etching SOI structure – SIMOX, sold epitaxial growth

Micro-lithography

Figure 3.10 shows the comparison between the exposure methods for optical X-ray, including synchrotron orbital radiation (SOR), and electron beams in terms of resolution and throughput. Optical exposure has been used mainly in the fabrication of VLSI. With a minimum feature

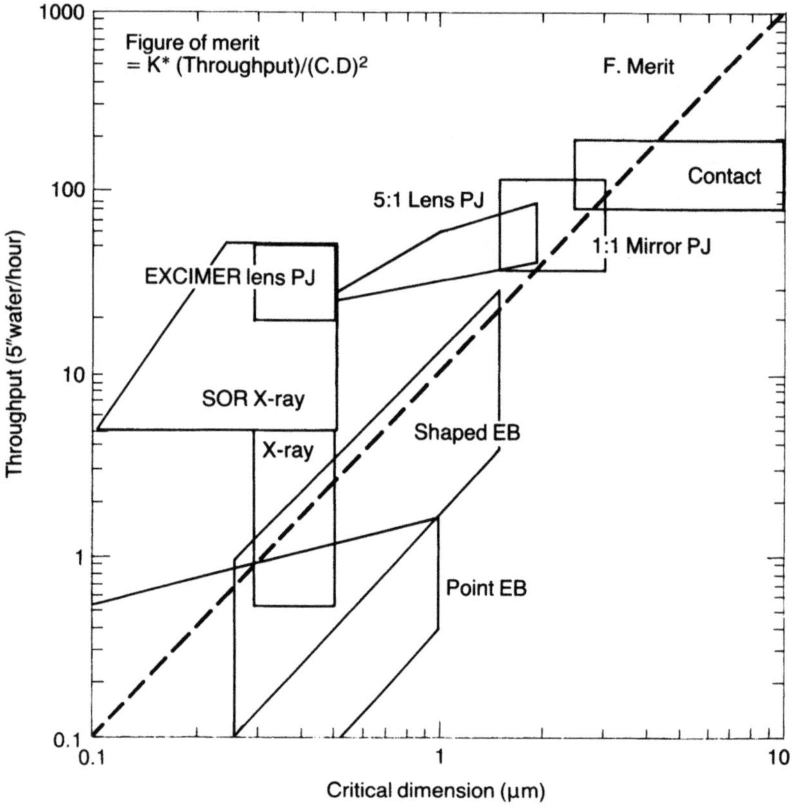

Figure 3.10 Comparison of micro-lithography technologies from the point of figure of merit.

size of 0.5 μm, for example, a resolution 0.4 μm is necessary to specify the conditions for exposure development and resist thickness. Moreover, the accuracy required for the critical dimension of the pattern and alignment should be at least 0.1 μm, and 0.15 μm respectively [3]. It is necessary for the accuracy to be much smaller than the fineness of the micro-lithography. For mass production of VLSIs, the minimum requirements on throughput and field size are about 20 wafers/h and 15 mm square, respectively. The optical exposure system has to satisfy these conditions. However, neither X-ray nor electron beams are capable of going below 0.5 μm. There is a trade-off between the accuracy and throughput as expressed by the following.

1. *Optical*: This has improved significantly and the resolution, R, of a projection aligner is calculated from the relation:

$$R = \frac{k\lambda}{NA}$$

where λ, and NA are the exposure wavelength and the aperture of the lens, respectively. The resolution of an existing optical system can be estimated, assuming $k = 0.8$. At present, it is usual to have a 5-to-1 reduction on the projected image based on the k-line (wave length = 436 nm) of a Hg lamp and a lens with $NA = 0.4$. To improve the resolution, there are two approaches which can be employed: shorter wavelength or a lens with a larger value of NA. For example, two lenses having $NA = 0.54$ for the g-line and 0.42 for the i-line (wavelength, $\lambda = 365$ nm) can be given a pattern of 0.8 μm. Furthermore, it may be possible to work to 0.5 μm if multi-level resist and contrast enhancement are used in conjunction with a high resolution lens.

Excimer lasers are now being employed as a short wavelength light source using an inert gas compound which is unstable at the ground state and so is self absorbing. A powerful pulse laser beam in the range of wavelength 150–300 nm is obtained by changing the gas composition. The KrF laser ($\lambda = 248$ nm) is the best optical aligner because the transmittance of the optical lens materials becomes smaller at shorter wavelengths. Patterns with a minimum size of 0.35 μm have been obtained with this.

The optical exposure system is only one of a number of lithographic technologies used on the production line. Compared with X-ray and electron beam systems, there is a higher throughput and a more complete set of support techniques and the excimer laser makes it possible to work to line widths of 0.35 μm. With the development of lenses with large values of NA, short wavelength light sources can be used to reduce the depth of focus. The operation also requires more precise planarization of the resist surface which imposes high levels of control on the operational variables. In practice, the resolution limit is determined by the permissible tolerances.

Miniaturization depends on the resist technology as well as the exposure. Multi-level resists are needed to generate very accurate patterns on stepped wafer surfaces or where there is back-scattering of the optical and electron beams. Simplification of the multi-level process is necessary because otherwise it is too complicated to use in practice.

2. *X-ray*: this method has concentrated on developing practical techniques of implementation which have required a critical evaluation of the lithography rather than improving the optical techniques, since the latter are approaching the theoretical limit of resolution. The problems are associated with deterioration in the depth of focus.

   Since X-rays have the advantage of increasing the depth of focus compared with the visible light, it is not necessary to use a multi-level resist. Nor is it affected by tiny particles. On the other hand, there are some problems, one of which is the accuracy of the mask. In many cases, misalignments are caused by disorders in the alignment of the masks. The masks are formed on a thin membrane of Si_3N_4 (several μm) and the patterns are drawn on a metal-absorbing film of Ta. The membrane is transparent to X-rays. To achieve a resolution of 0.35–0.2 μm with X-rays, the formation of the mask, including inspection and correction, must be looked at in some detail.

The throughput of X-ray system needs to be improved and to increase the intensity of the source, a plasma has to be used. Small plasma X-ray sources with gas-injection are available and are capable of increasing the throughput by a factor of 10 over that based on electron injection. Moreover synchrotron orbital radiation (SOR) is capable of increasing the throughput and can give a resolution of 0.1 μm because of the highly directional beam. The problem is the development of small and cheap equipment although some small prototype SOR rings are already under construction in several laboratories.

3. *Electron beam*: This approach uses a computer to draw the VLSI patterns directly. Essentially, it is a direct method which does not use a mask so it is different from micro-lithography based on the optical and X-ray exposure. It has also been used to make masks for both the optical and X-ray techniques and is close to working for fine resolution micro-lithography.

   The electron beam technique has several problems. At low throughput, a high speed electron beam machine with large angle deflection and a continuously moving stage is required. In the case of a minimum pattern size of 0.5 μm, the throughput is increased to about 20 wafers of 12.5 cm diameter (5 inches) per hour, but it should be noted that:

   (a) a correction procedure is necessary to compensate for fluctuations in the electron scattering in order to be able to draw accurate patterns on the wafer;
   (b) a stable treatment for the highly sensitive resist is required to achieve high throughput; and
   (c) some evaluation and compensation for the electrical damage caused by exposing devices to electron beam radiation is necessary.

   When these problems are overcome, electron beam direct drawing should be the best technique, especially for customized VLSI devices.

In addition, there are general problems with lithography associated with maintaining the alignment accuracy and the critical dimensions. Alignment accuracy relates to the positional accuracy of the patterns in the lower and upper layers. In order to reduce the size, or the memory for example, an increase in the alignment accuracy is essentially equivalent to reducing the minimum pattern size. This is difficult, although there have been several reports of patterns of less than 0.1 μm. It is rather more difficult to maintain accuracy of the critical dimensions by the use of micro-lithography. Future micro-lithography will have to overcome the very difficult problems in order to improve pattern accuracy.

Dry etching

Breakthroughs in other process technologies as well as lithography will be necessary to solve many of the problems in producing very high density devices. For sub-micron processes, it will not be sufficient to simply develop micro-lithography in terms of the scaling rule but will necessitate the introduction of new methods to achieve fine etching. Because fine patterns are required, it is necessary to reduce the amount of the undercutting during anisotropic etching to ensure high accuracy. To reduce malformation of the pattern the selectivity of the reactive gas with respect to the mask material also needs to be improved.

Fine patterning also requires a reduction not only in pattern sizes on the wafer surface but also the depth. This calls for shallower junctions from the impurity diffusion layers forming the transistors. However, there is a limited scope for improvement of the metallization process, so it is necessary to develop other approaches such as fabricating three-dimensional devices having a high aspect ratio and building structures with higher levels of integration. This, in turn, poses new problems. If a deep Si trench is formed using RIE (reactive ion etching), for example, the etching rate decreases as the etching pattern width becomes narrower (as in Fig. 3.11) and control becomes difficult, as shown in Fig. 3.12(a). Figure 3.12(b) shows a vertical cross-section through a Si trench using side-wall protection to prevent side etching. As the junction of the device becomes shallower, degradation of the device characteristics caused by ion radiation from plasma etching and dielectric breakdown from charge build up becomes critical.

The diameter of Si wafers presently used for VLSI is 15 cm (6 inches) but this is expected to increase to 20 cm (8 inches) in order to increase production rates. With such a increase in diameter, the manufacturing method will need to change from conventional batch operation to one-at-a-time processing. A chemical vapour deposition (CVD) reactor based on this mode of operation must be capable of high etching rates so as to match the proposed production rate.

In summary, the crucial factors in etching technology are achieving

high accuracy
high selectivity
precise control of the shape

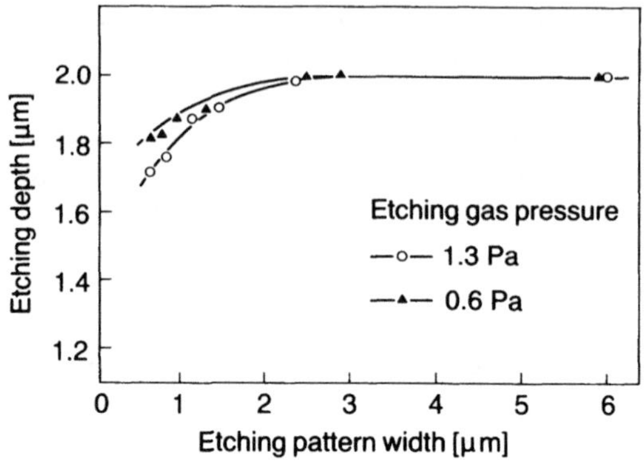

Figure 3.11 Dependence of etching depth on pattern width.

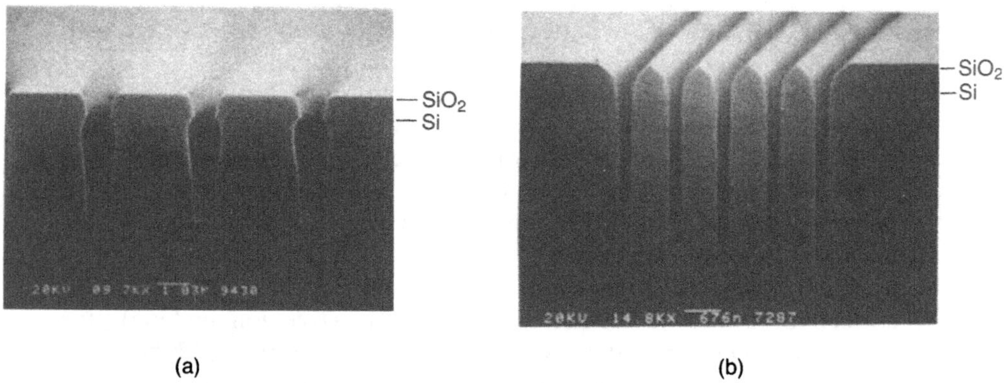

Figure 3.12 Cross-sectional etching profile of fine Si trench: (a) conventional RIE method; and (b) RIE method using side-wall protection.

minimum damage to the electrical properties of the devices
high etching rates and
improvement in consistency.

Trade-offs need to be made between some of the etching parameters in order to satisfy these requirements. In practice, there is only a limited set of options which will give the best values.

3.2 FUTURE DEVELOPMENTS

3.2.1 New materials

Silicon crystals will form the bulk of the applications in the future except for the group III–V compounds, as shown in Table 3.1. In addition, it is necessary to reduce crystal defects, to planarize Si mirror-polished wafers and to make Si wafers of larger diameter.

Si materials

Defect-free Si epitaxial wafers will find application in future MOS devices as designs are simplified. The bottle-neck is cost, which is almost three times higher than that of ordinary Si mirror-polished wafers. Reactors having a high throughput of the expitaxial wafers are being developed which can be expected to reduce the relative cost to less than a factor of 1.5.

Group III–V compounds

Vapour and liquid-phase growth, as well as molecular beam expitaxy (MBE) have been used to produce crystals. The first task is to find an optimum method of crystal growth which will give improved throughput.

3.2.2 New technology

Micro-lithography and etching

Higher levels of integration will depend on improvements in micro-lithography which must be realized in order to be able to mass produce devices. It is essential to make sure that at the same time, the device and system technologies are appropriately matched. The difficulties with micro-lithography will require further developments to be made to allow mass production of devices. At the same time, the basic unit and system technology will need to evolve.

Table 3.3. outlines the history and provides a summary of micro-lithography and techniques such as etching, film formation, and evaluation. Figures 3.13 and 3.14 [7] show some progress has been made in developing micro-lithography and plasma etching.

Several new approaches to RIE have been proposed which can be classified into two broad groups based on magnetron discharge and triode electrode construction. The former gives a high efficiency discharge over a range of low gas presures and results in rapid high accuracy etching with relatively little damage, because of the low self-biased voltage. An additional electrode is used to control the discharge which then independently adjusts the plasma density and hence the etching rate and ion energy. The latter affects the anisotropy and can damage the etched wafers. This simultaneously speeds up the etching while still maintaining high accuracy and good control over the patterns. The microwave discharge method employs a 2.54 GHz source while ECR (electron cyclotron resonance) is

Table 3.3 Trend of micro-fabrication process [4]

DRAM (bit)		*16* K	*64* K	*256* K	*1* M	*4* M	*16* M
		Minimum line width					
Process		5μm	*3–2.5* μm	*2–1.5* μm	*1.3–1.0* μm	*0.8–0.6* μm	*0.6–0.5* μm
Litho-graphy	Mask fabrication	Optical method		Electron beam method			
	Aligner	Contact aligner		1:1 Projection aligner			Excimer laser Electron beam X-ray
				5:1 / 10:1 Reduction projection aligner (g line → i line, shorter wave length)			
	Resist	Negative type		Positive type (→ multilayer resist → New-resist)			
Etching	Poly-Si Si_3N_4	Plasma etching		Reactive sputter etching → μ-wave etching. (μ-wave etching at low temperature)			
	SiO_2	Wet etching					
	Al						

Thin Film Formation	Metallization	Al–Si			Al · Cu · Si + Barrier metal, layer structure
	Interlayer	PSG, CVD · SiO_2			
	Passivation film			Plasma SiN, plasma SiO	
				Planarization (spin or glass, bias-sputtered SiO_2 TEOS oxide[a])	
Measurement	CD measurement	Optical method			
					Scanning electron microscopy method
	Dust	Optical method	Laser method (dust monitoring) (water defect inspection)		Scanning electron microscopy method
	Particulate	Optical method	Laser method (scattered light, polarized light method)		

[a] TEOS: $Si(OC_2H_5)_4$, Tetraethoxy Silane

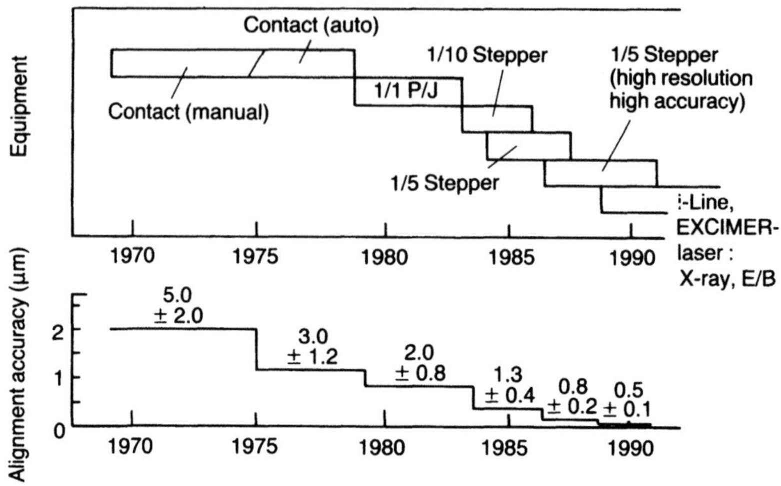

Figure 3.13 Progress in micro-lithography equipment [7].

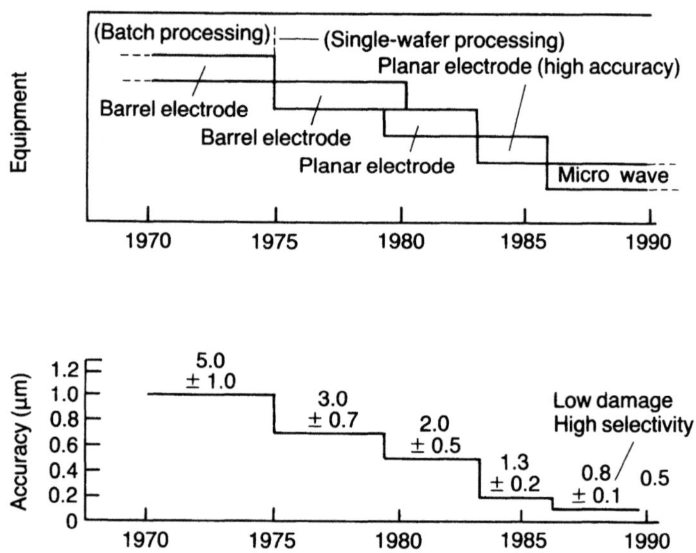

Figure 3.14 Progress in plasma etching equipment [7].

based on a magnetic field. ECR increases the discharge efficiency by working at low gas pressures. This method is different from conventional RIE because the ECR equipment has wafer-etching and plasma generation chambers. It is possible to install an electrode between the two

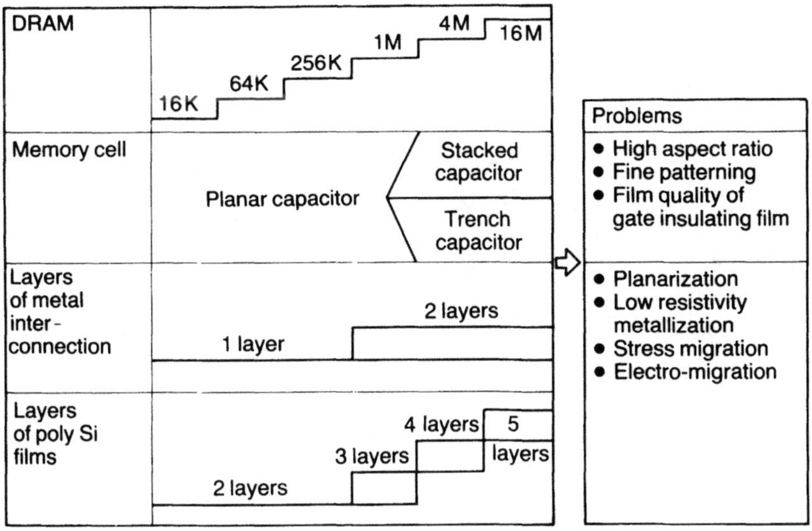

Figure 3.15 Evolution of multilevel metallization.

chambers or add a bias voltage to a substrate electrode in order to control the energy of the injected ions. This is the so-called RIBE (reactive ion beam etching) process.

All the above-mentioned methods are essentially variations of the conventional RIE in that they use a plasma. Radiation damage arises from collisions between high energy particles. One method of overcoming such problems is to use photo-excitation etching.

Multi-level metallization

With VLSI, it is essential to reduce the access time of external signals to the cells so as to maintain high chip speeds. Multi-level metallization increases the operating speed and uses a smaller Si chip area. Figure 3.15 shows the trend in multi-level metallization in MOS memories used for DRAMs. Al, poly-Si and refractory metal silicides such as WSi_2 and $MoSi_2$ have all been developed for use in devices. The problems of fine multi-level metallization are indicated in Fig. 3.16.

Fine multi-level metallization makes use of methods developed for metal and insulation film formation, micro-lithography and similar techniques. It is also necessary to find methods for achieving planarization and low contact resistance interconnection multi-level metallization while at the same time improving the materials. To increase the reliability of fine metallization, problems of electro- and stress migration and Al

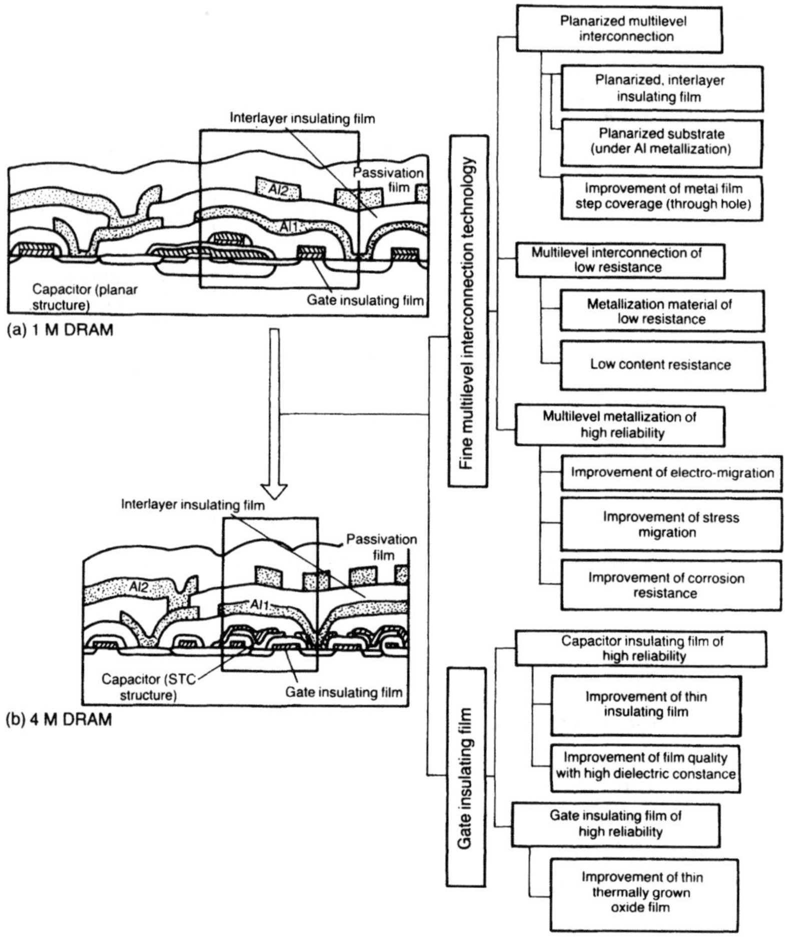

Figure 3.16 Problems of fine multilevel interconnection and gate insulating film technology in applications for megabit DRAM [4].

corrosion need to solved. In particular, electro-migration means that conductive materials need to be transported by high density electron flow to ensure fine metallization.

Analysis and evaluation

With advances in fine micro-lithography, analysis and evaluation techniques also need to be developed. The important diagnostic methods are:

1. high sensitivity physical instrumentation to probe very small areas, including scanning and transmission electron microscopy, secondary

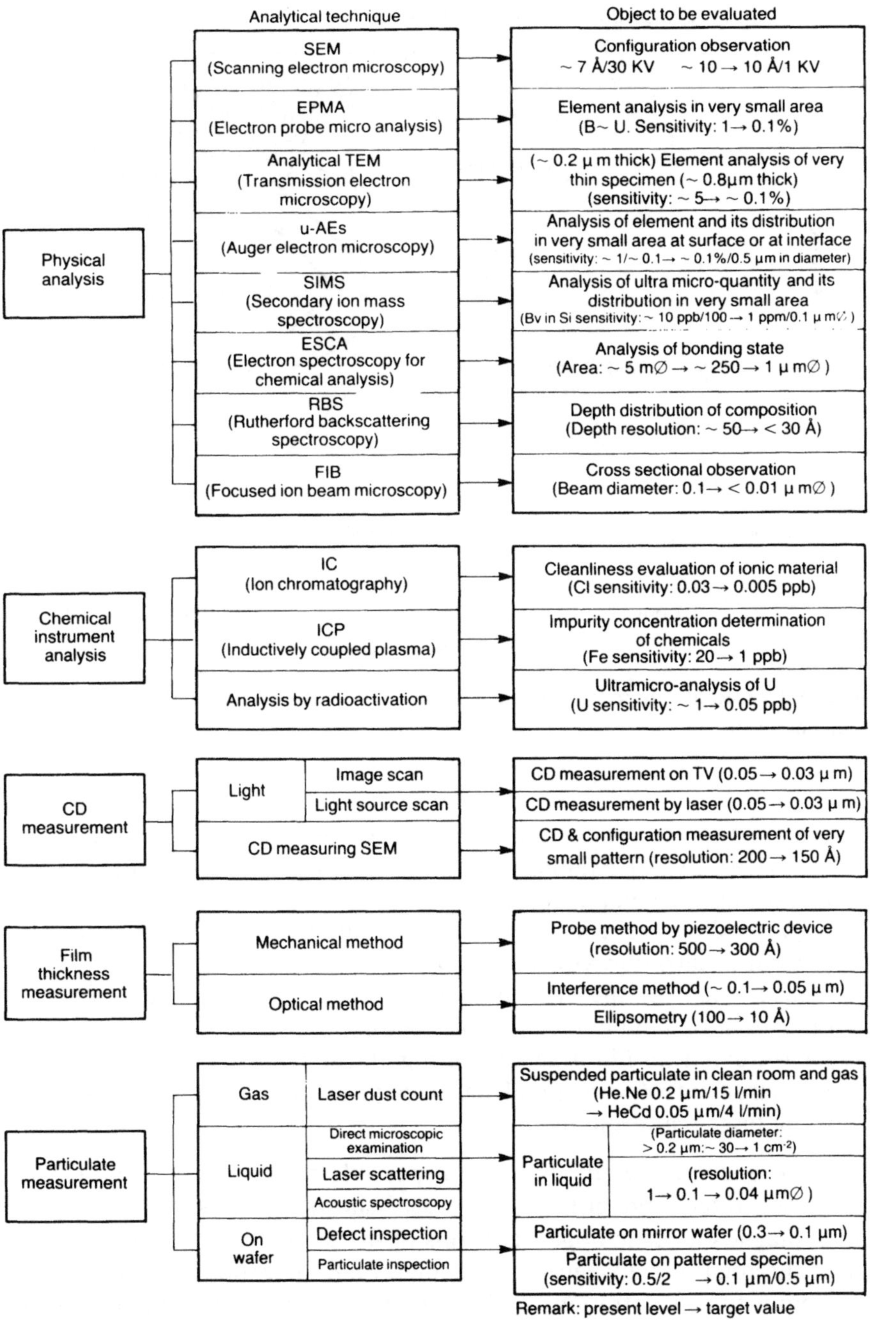

Figure 3.17 Main diagnostic technology in semiconductor processes.

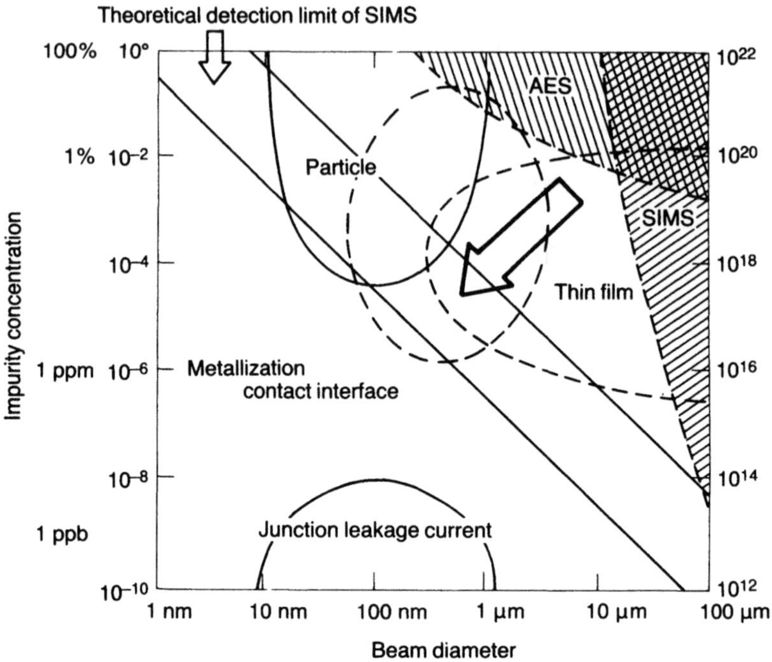

Figure 3.18 Beam diameter and detection limit required for analysis of impurity concentration.

ion mass spectroscopy, and micro Auger electron spectroscopy – these are necessary for examining the morphology and crystal defects and measuring the composition;

2. high sensitivity chemical analyses for uranium and other impurities present in ultra-low concentrations;
3. measurement of the critical dimensions of very fine structures using optical and scanning electron microscopes;
4. measurement of film thickness using ellipsometory, etc.; and
5. measurement of very small particles in gas or liquid or on Si wafers using laser scattering methods.

These analytical methods make it possible to characterize the devices more accurately and faster. Figure 3.17 shows the present status of the main diagnostic technologies in terms of the target values, while Fig. 3.18 gives the relation between beam diameter and detection limit in monitoring impurity concentrations. It is clear that there is a continuing demand for detection techniques for increasingly lower levels of impurities.

More insight into the relationships between operating conditions, reaction mechanisms during film formation and etching and characteristics

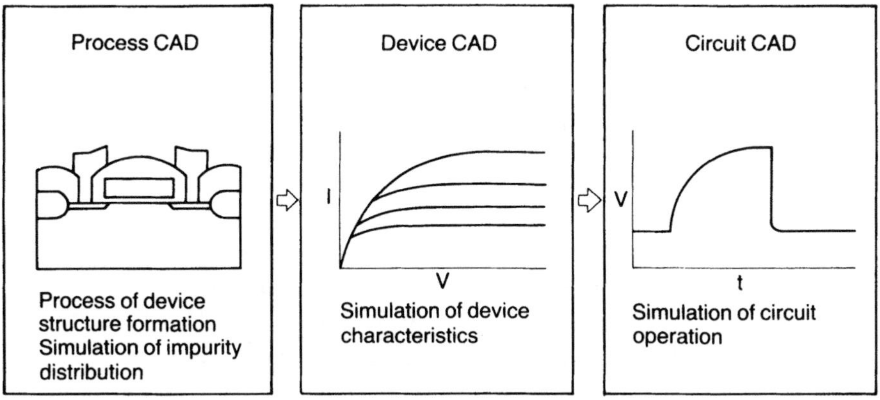

Software name	Dimension 1D	Dimension 2D	Dimension 3D	Impurity	Shape
Stanford Univ.: SUPREM	O			O	
Stanford Univ.: SUPRA		O		O	
Calif Univ.: SAMPLE		O			O
Hitachi: SPIRIT		O	O	O	O

Software name	Dimension 1D	Dimension 2D	Dimension 3D	MOS	BIP
Hitachi: CADDETT		O		O	B O
Hitachi: CADDETH			O	O	O
Stanford Univ.: GEMINI		O		O	
Waterloo Univ.: WATMOS			O	O	
IBM: FIELDAY			O	O	O
NTT: TRANAL 3D			O	O	O
Aachen Univ.: MEDUSA		O	Pseudo O	O	O

Software name	DC	AC	Transient
Calif Univ.: SPICE	O	O	O
Hitachi: HSPICE	O	O	O
Aachen Univ.: MEDUSA	O	O	O

Figure 3.19 Concept of total simulation. SUPREM: Stanford University Process Engineering Model [95]; SUPRA: Stanford University Process Analysis [96]; SAMPLE: Simulation and Modeling of Profiles for Lithography and Etching [97]; SPIRIT: Simulation Processor for Integrated Representation of Impurity-profile and Topography [98]; CADDET: Computer Aided Device Design in Two dimensions [99]; CADDETH: Computer Aided Device Design in Three dimensions [100]; GEMINI: [101]; WATMOS: [102]; FIELDAY: Finite Element Device Analysis Program [103]; TRANAL: Transistor Analysis Program [104]; MEDUSA: [105]; SPICE: Simulation Program with Integrated Circuit Emphasis [106]; HSPICE: High-speed SPICE [107].

of products is necessary to improve the throughput while ensuring the reliability of the devices. Chemical engineering methods are well-suited to this important role.

For highly integrated devices, it is essential to be able to predict the

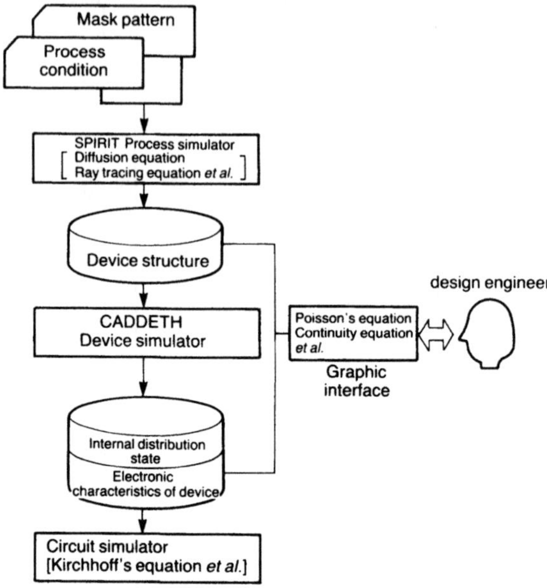

Figure 3.20 System construction of total simulation, showing the simulation system which consists of process, device and circuit simulation: SPIRIT = simulation processor for integrated representation of impurity-profile and topography; CADDETH = computer aided device design in three dimensions; [#] denotes the equation used for the simulation.

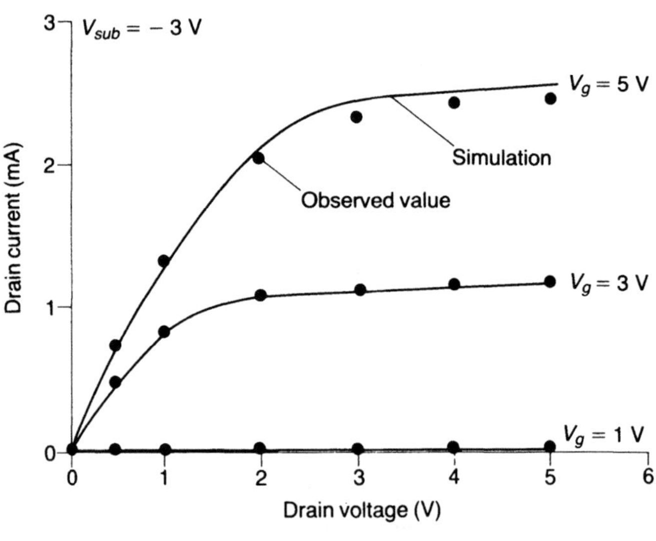

Figure 3.21 *V–I* characteristics of MOS transistors, showing static characteristics of transistor obtained by total simulation.

overall behaviour of components from knowledge of the basic elements. For example, a memory cell is a basic element of a memory device. It is possible to simulate the performance of a cell. Figure 3.19 lists some packages capable of carrying out simulations which can be used to improve manufacture of VLSI chips. They provide detailed information on the performance and this shortens the time to make samples. Figure 3.20 illustrates how the simulation packages are related [8]. The two-dimensional process simulator, SPIRIT (simulation processor for integrated representation of impurity-profile and topography) can analyse the complicated device shapes as well as how the distribution of impurities influences performance. The current and capacitance characteristics can then be calculated using a three-dimensional device simulator, CADDETH (computer aided device design in three dimensions). As shown in Fig. 3.21, this predicts the relationship between the drain current and voltage in a MOS transistor and so can be used in circuit simulation. The ultimate goal of such devices is to emulate the human brain, which has of the order of 14 billion cells. Current VLSI devices have very much less than this but there is considerable interest in neural networks, and this has given rise to many new developments. However, while some progress has been made, there is a limit to what can be achieved.

4

Characteristics of VLSI

VLSI (very large scale integration) refers to devices based on transistors, diodes, resistances and capacitors assembled at very high densities. In this chapter, the principal features of such devices are discussed, together with the characteristics of the electronic circuits.

4.1 DEVICE STRUCTURES

In general, there are two types of elements involved in electronic circuits: active and passive. A passive element does not add energy to a signal but it can give rise to bias (diodes, resistances, capacitors, inductors are typical passive elements). Active elements add energy to a signal as, for example, in the transistor. Since the structure is complicated, it is called a 'device'. Device is not really the most appropriate word, any more than 'element' is for an integrated circuit (IC). It is best thought of as a generic term. In fact, 'device' and 'element' are used interchangeably. VLSI devices should be seen as complex electric circuit devices composed of active and passive elements.

The important active devices in VLSI devices are transistors. There are two types: bipolar and FET (field effect transistor). Bipolar has two types of carrier namely positive holes and electrons, while the FET is based on either positive holes or electrons so is sometimes referred to as a monopolar transistor because the resistance path can be changed by an external electric field. Bipolar transistors can switch several times faster than FETs so they are widely used in equipment which requires fast response. Unfortunately, they have a relatively large power consumption.

It is appropriate to explain the structure of transistors and passive elements so that the way in which they work can be understood. At the highest densities, the line widths for the circuits are less than 0.1 µm. On this scale it is necessary to account for the wave property of the electrons when looking at the characteristics.

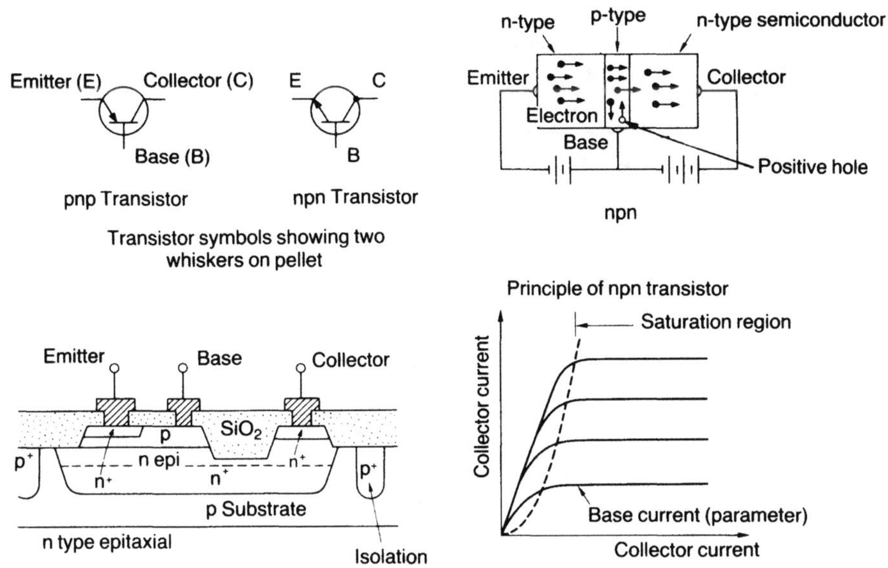

Figure 4.1 Fundamental characteristics of bipolar transistors.

4.1.1 Bipolar transistor

There are two basic semiconductor material structures for bipolar transistors: n-type based on electrons and p-type based on positive holes. Both can be assembled in the form of a sandwich of n-p-n or p-n-p to form transistors. The bipolar transistor was developed before the FET. Its structure is more complicated than that of the FET but the response speed is much faster. The principle on which it works is the control of current from an emitter to collector using a very small current added at the base. In the case of an n-p-n transistor, there are two currents to the base, an electron current from the emitter and positive hole current from the collector. Because of the sensitivity to changes of current flow through the transistor, it is possible to control a large current by adding a few positive holes to the base after which they recombine with electrons. For the p-n-p transistor, the reverse effect is observed, i.e. electrons are used to neutralize the positive holes.

Figure 4.1 shows the basic symbols and characteristics of a bipolar transistor and a cross-section of a typical n-p-n transistor in a VLSI device. Compared with the p-n-p transistor, where the electron mobility (the ratio of electron speed to the strength of the electric field) is several times larger than the hole mobility, the n-p-n form acts faster. The

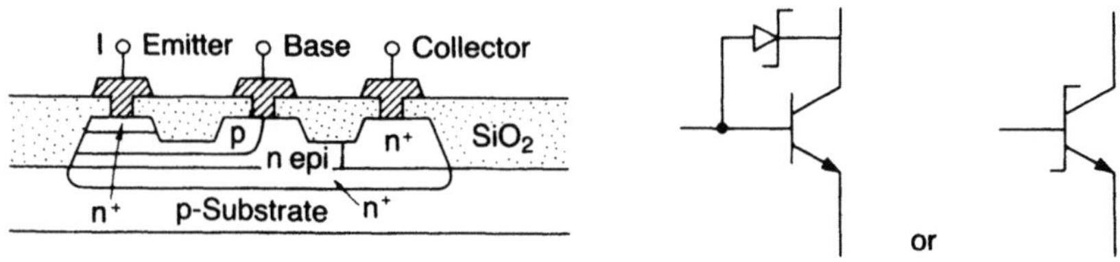

Figure 4.2 Schottky clamp n-p-n transistor and symbols used.

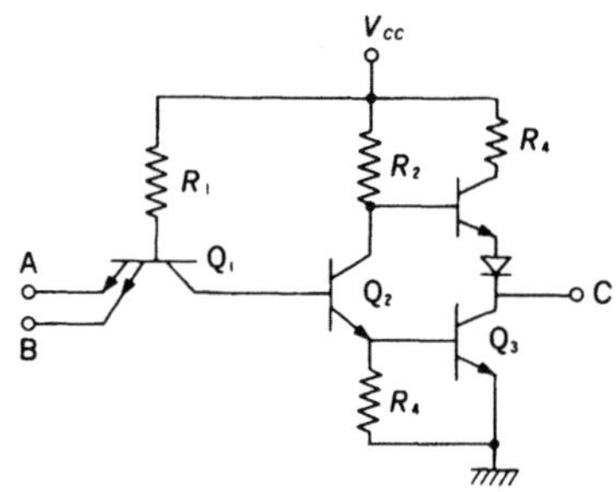

Figure 4.5 IIL circuit cross-section and symbol used.

Schottky barrier transistor has a higher response speed and is a kind of bipolar transistor. This is achieved by adding a Schottky barrier diode between the base of an n-p-n transistor and the collector. The diode then works as a clamp diode so as not to saturate the transistor action which results in a high speed response (Fig. 4.2).

Bipolar transistors are used for high speed logic circuits. Examples are ECL (emitter coupled logic), TTL (transistor-transistor logic), CML (current mode logic) and IIL (integrated injection logic) devices. A TTL device uses a transistor in the saturated state. Figure 4.3 shows a NAND circuit using a multi-emitter transistor in the input circuit. When signals are input to A and B simultaneously (logical value 1) transistor Q1 is activated so that Q2 operates Q3 which is then activated by the current to R4 so the output signal voltage from Q3 is zero (logical value 0). On the other hand, when both A and B, or either A or B are zero, Q1 and Q2 do not work. Because the current to the collector Q2 increases, Q4 and the diode are activated and the output signal voltage of the terminal C is no longer zero (logical value 1).

An ECL device is a logic circuit which has a common emitter. It can reduce an input signal of 3.5 V from a TTL circuit to 0.8 V and operates at high speed. Figure 4.4 shows a NOR/OR circuit. When the signal voltage is lower than the reference voltage Vr, Q3 is activated and increases the current in R2. Consequently, because the voltage of the

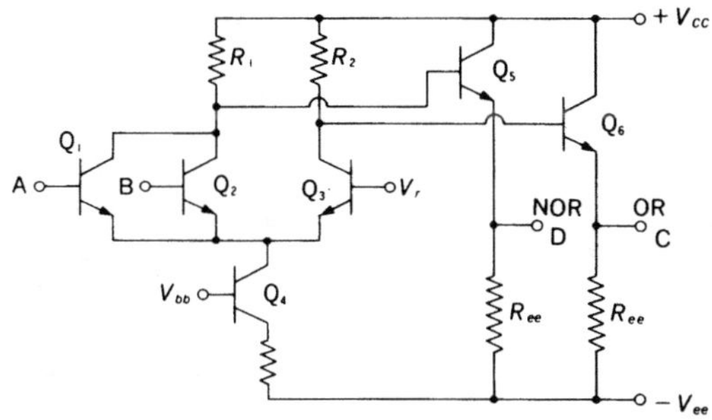

Figure 4.4 NOR/OR gate using ECL.

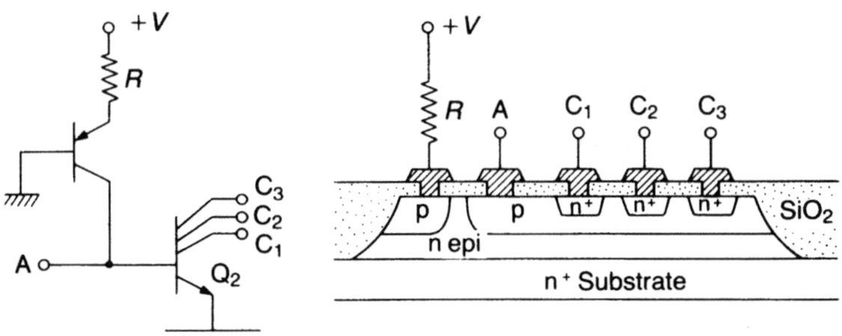

Figure 4.5 IIL circuit cross-section and symbol used.

collector Q3 and emitter load R3 of Q6 decrease, the logical value takes a value of 0 at the output of the OR circuit. When the voltage of either input increases, Q3 assumes an off-state and the logical value at the output becomes 1. A similar analysis shows that for the NOR circuit the output state is opposite to the OR circuit. With this arrangement, since Q5 and Q6 are always active, power consumption is high. CML therefore corresponds to an ECL circuit without Q5 and Q6.

As shown in Fig. 4.5, the IIL has a structure where the base of p-n-p transistor works as an emitter in a n-p-n transistor as well as on its own account. If the input terminal takes a high voltage (logical value 1), the n-p-n transistor is activated and the output decreases (logical value 0). Similarly, if the input voltage is low (logical value 0), current flows to Q1 and Q2 is not activated, so the output voltage increases (logical value 1).

4.1.2 MOSFET

Another important transistor is the metal oxide semiconductor field effect transistor, MOSFET. It comprises a current source, current drain and

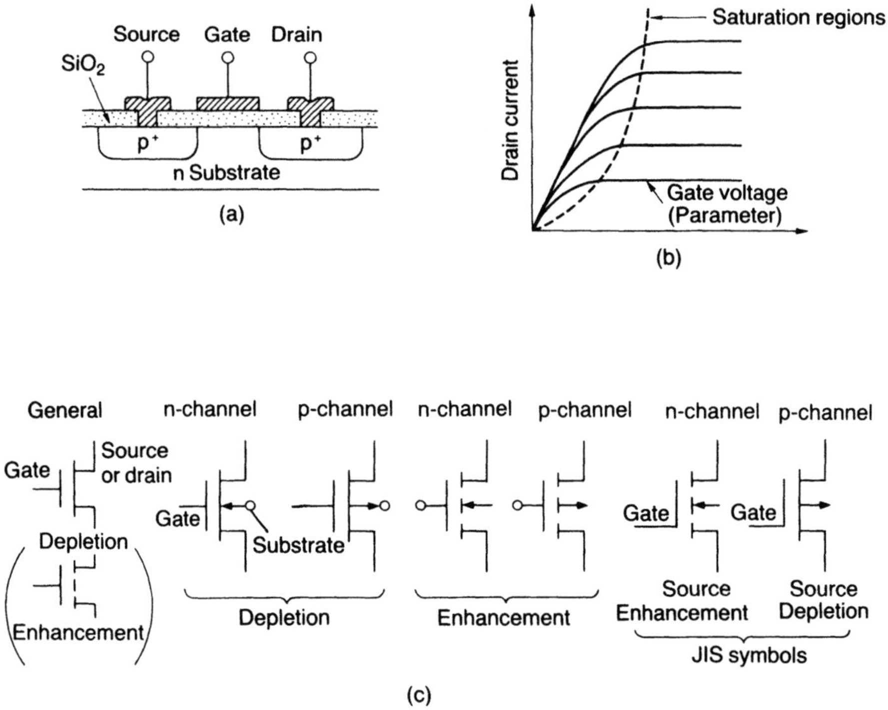

Figure 4.6 Fundamental features of MOS FET: (a) section of CMOS; (b) output characteristic of MOS; (c) symbols used.

control gate. The silicon wafer between the current source and drain is covered with silicon dioxide and a metal gate is formed on it (Fig. 4.6(a)). In the case of an n-type silicon wafer, the source and drain are p-type and a current path for the positive holes through the silicon dioxide should be a p-channel. For a silicon wafer of p-type, the source and drain should be n-type and the current path should be an n-channel. The relationships between drain voltage and current are shown in Fig. 4.6. It can be seen that as the drain voltage increases, the current increases and is bounded at the upper limit by the gate voltage.

There are two types of MOSFET: enhanced and depleted. In the case of the former, if the gate voltage is zero, there is no current, but if a threshold voltage is added to the gate, a current flows. For the depleted MOSFET, if the gate voltage is zero, current always flows: if a voltage is added to the gate, the current stops. p-MOS and n-MOS are examples of enhancement and depletion, respectively. These can be reversed by controlling the impurity concentrations using ion implantation. The distinction between n-MOS and p-MOS, and between enhancement and

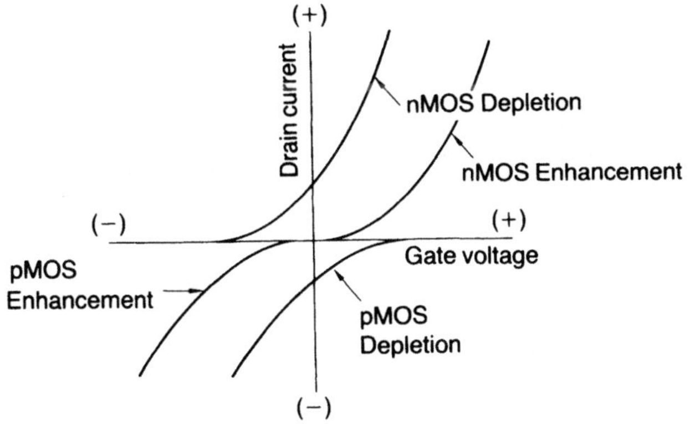

Figure 4.7 Drain current vs. gate voltage for MOS FET.

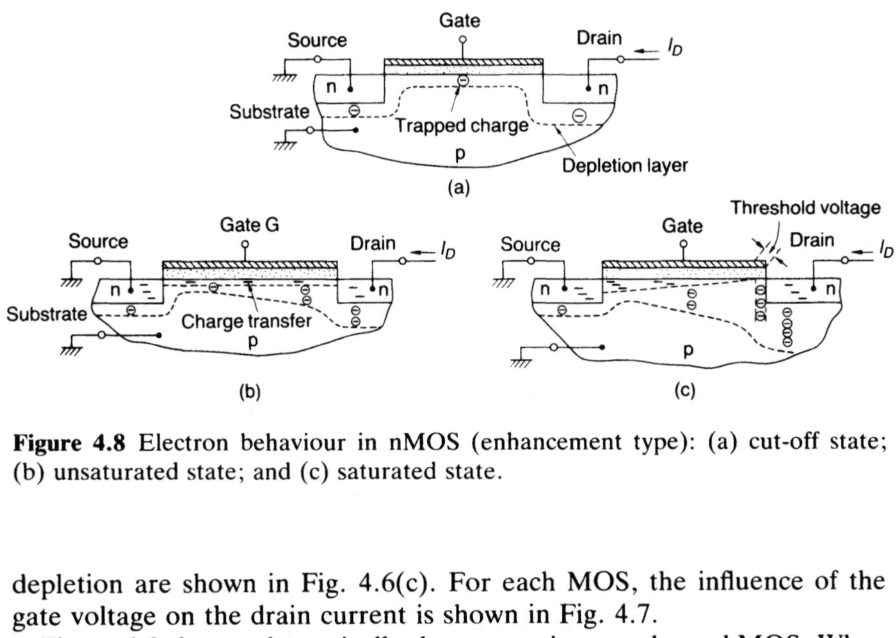

Figure 4.8 Electron behaviour in nMOS (enhancement type): (a) cut-off state; (b) unsaturated state; and (c) saturated state.

depletion are shown in Fig. 4.6(c). For each MOS, the influence of the gate voltage on the drain current is shown in Fig. 4.7.

Figure 4.8 shows schematically the current in an n-channel MOS. When there is no gate voltage, the voltage difference between source and drain is reverse biased and there is a depletion layer having no carrier electrons. Consequently, there is no current between the source and the drain. If there is positive voltage exceeding the threshold added to the gate, electrons are pulled to the boundary of the silicon dioxide layer, including the reverse layer, which then generates a current to the transistor. The action is controlled by using the gate voltage.

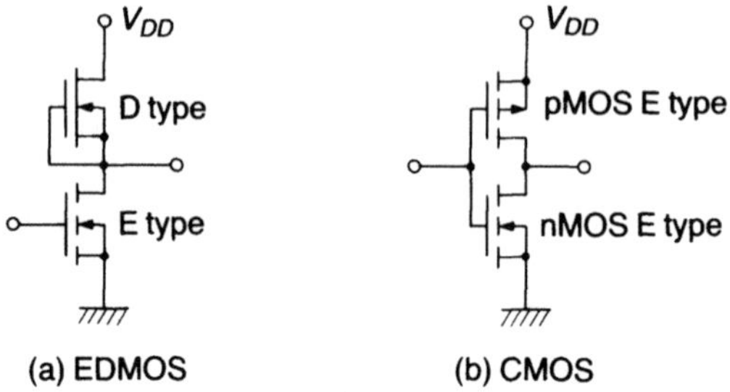

Figure 4.9 (a) EDMOS and (b) CMOS circuits.

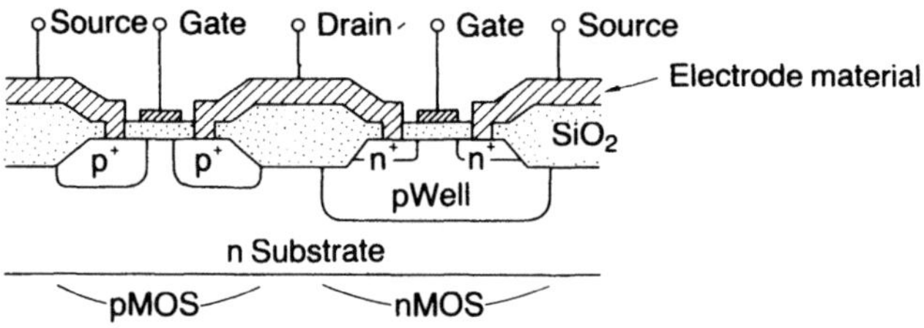

Figure 4.10 Cross-section of CMOS circuit.

The current in the MOSFET is generated by adding a voltage to the gate from a silicon dioxide insulating film. The input resistance of the direct current is greater than $10^{12}\,\Omega$ so a direct current is not needed in the on/off state. By combining the depletion and enhancement-type MOSFETs, a highly integrated very high speed (EDMOS) circuit is formed. The combination of p-MOS and n-MOS produces a complementary circuit (CMOS) which has high noise immunity and low power consumption. Figures 4.9 and 4.10 show such combinations and present a view of a cross-section through a CMOS.

4.1.3 Comparative characteristics

A device is characterized in terms of its power consumption and switching speed. In general, the important feature is the relationship between power consumption and switching delay time (the inverse of the switching speed). Figure 4.11 indicates the relationship between them, including the GaAsFET and Josephson device, which will be discussed later.

4.1.4 Scaling rules

A proportional scaling law was proposed by R. H. Dennard in 1979 which emphasized the benefit of using smaller transistors and hence the value of

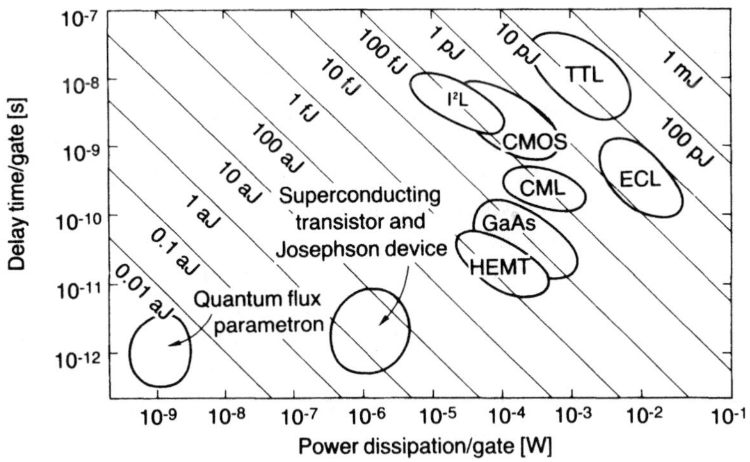

Figure 4.11 Relation between delay time and power dissipation.

VLSI. Reducing the size of the transistor makes it possible to have higher switching speeds and achieve circuit integration without functionality being lost. This law is especially useful in the case of MOS devices.

According to the scaling law, if the device size (length, thickness, width) of a MOSFET, is decreased by a factor of $1/k$, the concentration of the impurity should be increased by k times so the voltage and current are in the ratio $1:k$. Thus, though the electric density does not change, both device area and power consumption are reduced by $1/k^2$ and the electric capacitance by $1/k$, which therefore increases the switching speed by a factor of k. These relationships are acceptable for a line width of approximately 3 μm but several important problems appear for narrower lines. An example is the short channel effect which arises when the distance between the gate and drain is so small that the electric charge in the channel is influenced by the depletion layer induced by the electric field in the drain. Then, the threshold voltage decreases compared with the normal voltage. In the extreme case, a short circuit occurs between the source and the drain. The channel length should therefore be at least twice the thickness of the depletion layer for normal device action. As with other physical limitations, when the gate insulation film becomes too thin, a tunnel current will flow and if the impurity concentration at the surface is more than $10^{19}\,cm^{-3}$, the insulation film may break down as a result of the tunnelling current (tunnel break).

4.2 MEMORY TYPES

Using the convention introduced by the scaling law, the degree of circuit integration regularly increased by a factor of 4 every three years in the early stages of IC production: more recently, this has increased to a factor of 4 every four years. There are five kinds of semiconductor memory: DRAM (dynamic random access memory), SRAM (static RAM), ROM (read only memory), EPROM (erasable programmable ROM) and E^2PROM (electrically erasable PROM). SRAM uses a flip-flop circuit as the memory cell and does not need to be refreshed. DRAM is a memory cell formed by a silicon wafer and silicon oxide which can hold an electric charge for a limited time and so acts as a temporary capacitor. It is possible to assemble an extremely large number of such circuits using this simple structure. However, current leaks from the capacitor so the memory contents need to be refreshed periodically. With RAM, it is possible to both store in and read from the memory. With ROM, it is only possible to read data and it is the mask used in making the ROM which determines the contents.

4.2.1 ROM

Examples of a ROM mask are the use of fixed memory storage for well-established programs and storing special characters such as kanji (Chinese characters) in a dictionaries need in Japanese word processors. Reduction in the size of the memory cell is possible by using ion implantation or diffusion masks, for example in n-MOS or CMOS. With EPROM, memory data is stored electrically and can be deleted using ultraviolet light. This makes it suitable for program and data memory in microprocessors. It is an example of a structure where there is self-alignment of the control and floating gates. There are two types of

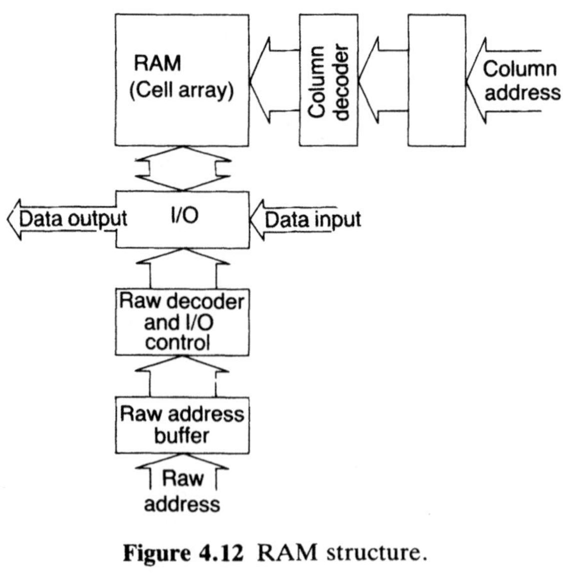

Figure 4.12 RAM structure.

E^2PROM. One utilizes a surface boundary trap of MNOS (Metal–Si_3N_4–SiO_2–Si) and the other utilizes tunnelling electrons supplied from a floating gate. Stored memory data is retained, even if the power supply is cut off.

4.2.2 RAM

RAM is used for temporary storage in memory. SRAM does not need a sophisticated circuit to refresh it but because the cell is larger than that for DRAM, it is difficult to make highly integrated circuits. Figure 4.12 shows a schematic diagram of the signal flow in a RAM. A single data item is stored and read by pointing to a column and row address. Figure 4.13 shows circuits for both n-MOS and CMOS SRAMs. The bit line (BL) is used for storing and reading data and the word line (WL) for addressing cells. In order to write data in a SRAM, the voltage of the word line is increased so that Q1 is activated. The datum and complement are written in S1 through BL and in S2 through $\overline{BL}$, respectively. The bit line is then isolated by switching WL off. For logic 1, Q2 and Q1 assume on and off states respectively and the state is maintained at V_{cc}. For logic 0, the inverse is obtained. In the case of logic 1, reading is executed by operating the bit lines, i.e. by making the voltages of BL and $\overline{BL}$, higher and lower, respectively. The values of the signals from BL and $\overline{BL}$ are then converted to digital form.

The cell structure of the DRAM is very simple, as can be seen from Fig. 4.14. It should be noted that n-MOS and CMOS SRAMs only require 1/4 and 1/6 the number of transistors. Accordingly, the cell area is proportionately smaller, so it is possible to obtain a highly integrated device having a large number of cells. In writing data from a transistor memory cell to a word, a line is activated and electrons charge a capacitor up to the voltage corresponding to the word line. In this case, the presence of the electrons corresponds to logic 1. In reading, the bit line is activated and the voltage of the word line increases. Because the charge electrons disappear in a time varying from several hundred milliseconds to a second, the stored data needs to be refreshed. Hence, the original datum should be read and restored at a frequency of tens of kHz. Capacitors for memory cells of less than 256K bits are made on the surface of the device, using an insulator. For Mbit memories, a stack or trench type capacitor is used, as shown in Fig. 4.15.

4.3 ARRANGEMENT OF VLSI DEVICES

4.3.1 Microcomputers

Semiconductor memory makes it feasible to design computers based only on semiconductors and thus to replace the magnetic materials previously

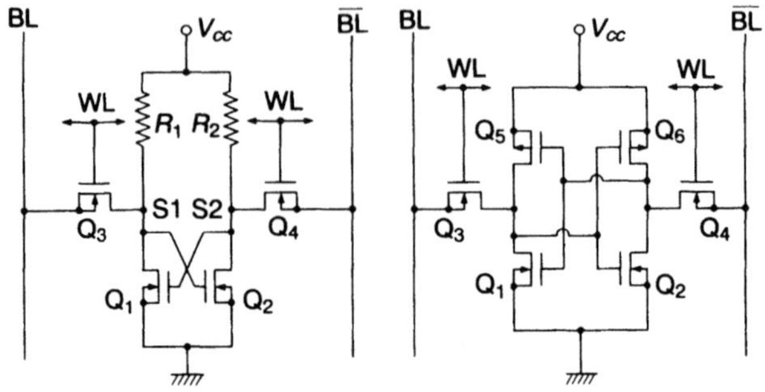

(a) n MOS SRAM cell (b) CMOS SRAM cell

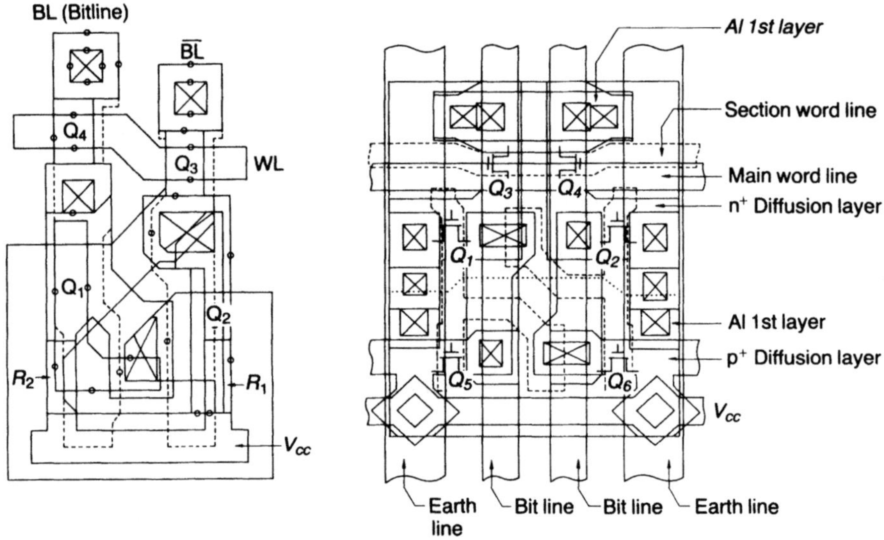

(c) n MOS SRAM Layout pattern *(d) CMOS SRAM layout pattern*

Figure 4.13 SRAM cells in n-MOS and CMOS.

used. This resulted in the Intel 4004 micro-computer in 1971 using a 4-bit word length comprising several thousand transistors. This can be compared with the 8-bit micro-computer which has about 10 thousand transistors and the 16 bit micro-computer requiring 50 to 100 thousand. The 32 bit micro-computers have several hundred thousand transistors.

Figure 4.16 shows a schematic functional diagram of a micro-computer. A clock is used to provide synchronization for transfer of data. The figure illustrates a processor based on an 8-bit word length and has an arithmetic

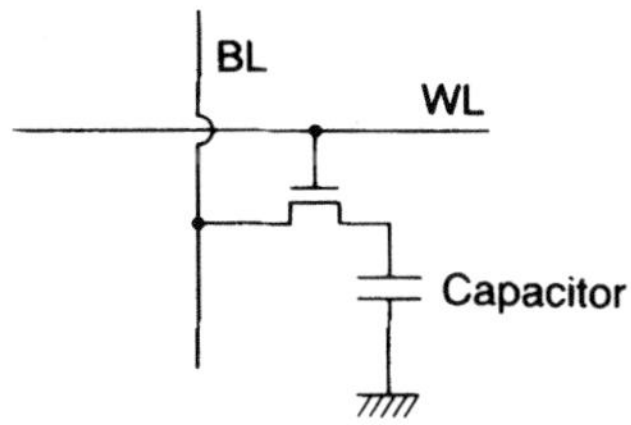

Figure 4.14 DRAM cell.

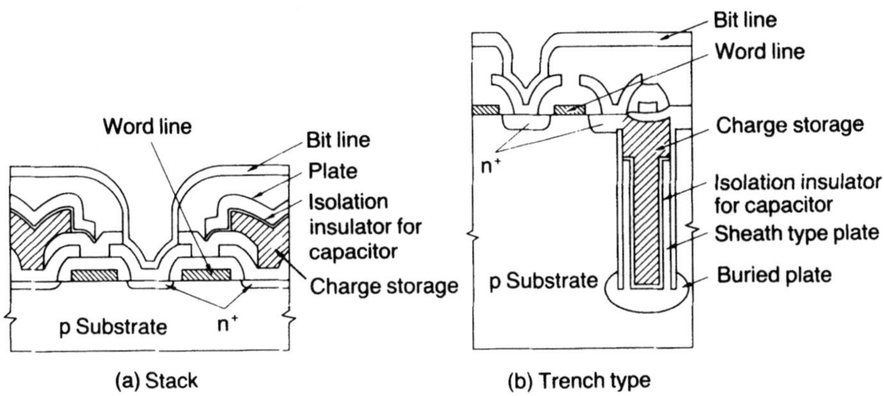

Figure 4.15 Cell structure in Mbit class DRAM terminal.

and logic unit (ALU), accumulator, register and memory data register (MDR). The number marked on the lines in the figure denotes the word length. The memory address is 16 bits, so it is possible to define 2^{16} (= 65 536) memory addresses. The address stack has a capacity of 7 levels of 16 bits. The instruction register has 24 bits: 8 bits for the operator and 16 for the operand. Control of the data-bus is performed by using a ROM which is micro-programmed with 128 words of 30 bits. If the memory capacity is insufficient, a PLA (programmable logic array) is substituted for the ROM or a subroutine program is used. Address control for the ROM is achieved with a bit address register.

Data transfer takes place between the I/O unit and CPU and the operations are all executed in the ALU, accumulator and address register using flags. The memory data register transfers data (bidirectionally) between the main memory and CPU. Data in the operand portion of the instruction register is transferred into the memory address register, which selects the address of the main memory. Instructions are transferred from the instruction register under the control of the ROM address and the

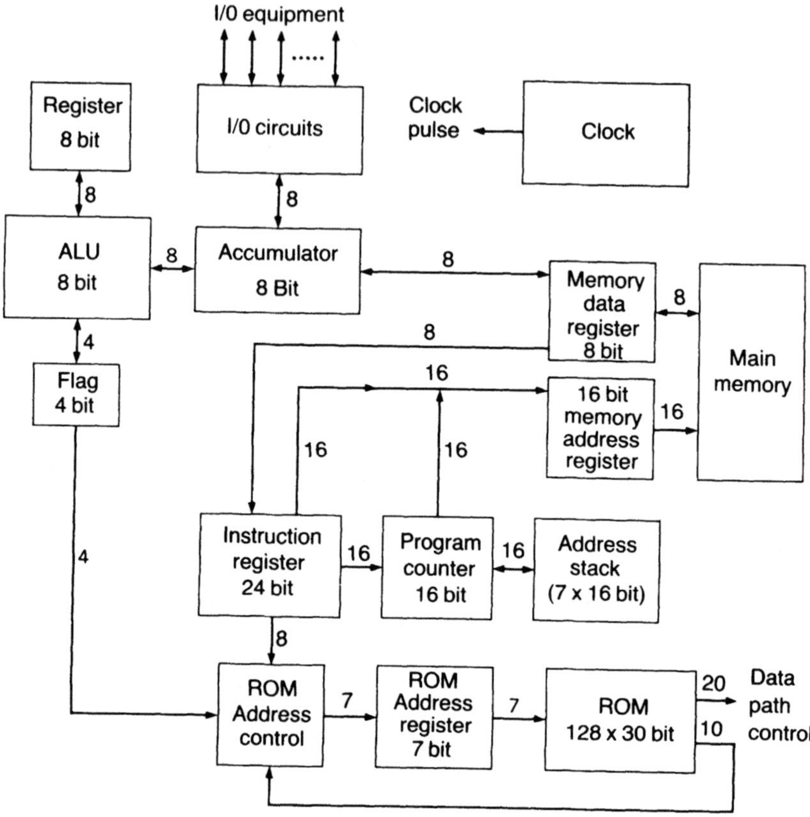

Figure 4.16 Structure of 8-bit microcomputer [9].

address control circuit. The ROM address control combines a flag, operand from the instruction register and the address from the ROM address register and points to the ROM address where the micro-instructions are stored. The CPU data bus, address for accessing main memory and activation of the ALU and I/O circuits are controlled by the data-bus control signal from the ROM. Figure 4.17 shows a modern 32-bit chip micro-computer comprising 730 thousand transistors.

4.3.2 Special purpose components

Mass produced VLSI components can be used to perform many functions but are not well-suited for high speed operation or very special functions where specific characteristics are desired. There is a need to develop a VLSI device having specific capability at low cost. To meet this demand PLA/PLD (programmable logic array/device) and gate arrays have been

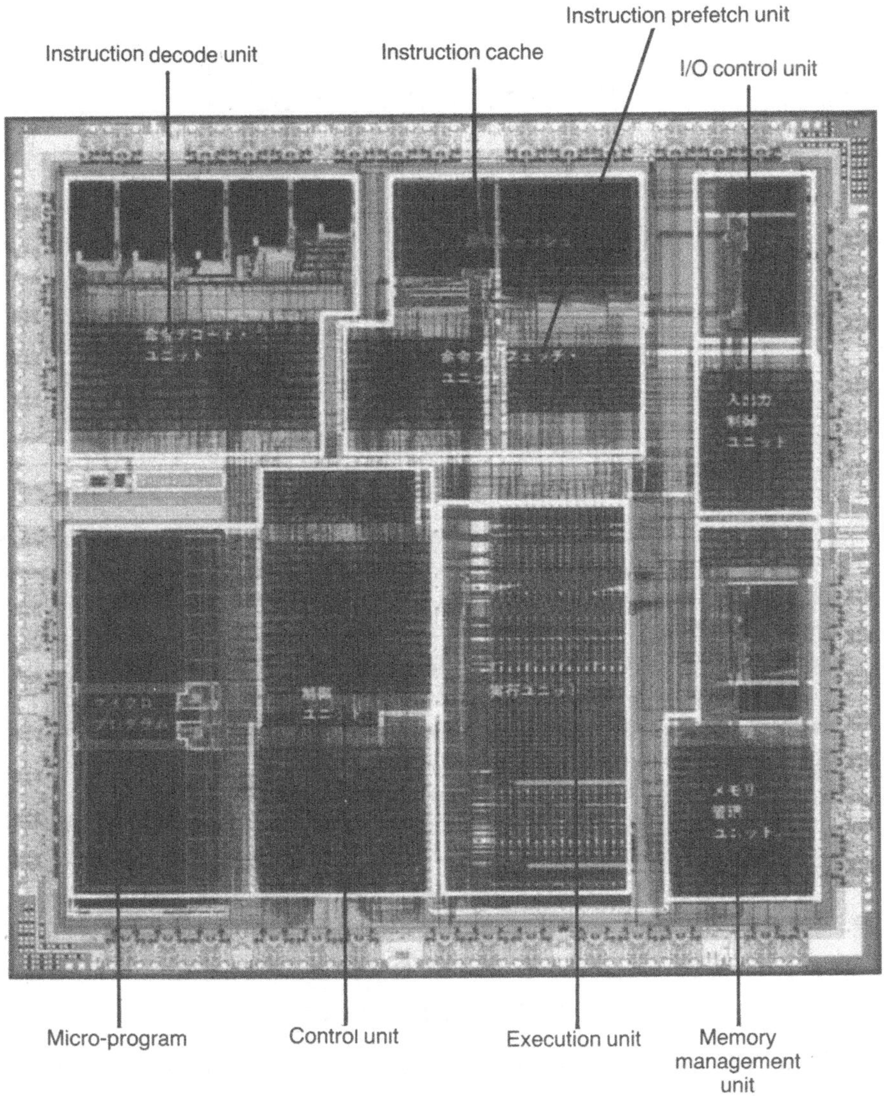

Figure 4.17 Top row of modern microcomputer chip is G micro/200. 730 000 transistors are integrated on 14.00 × 14.1 mm^2 chip using 1 μm design rule and CMOS technology.

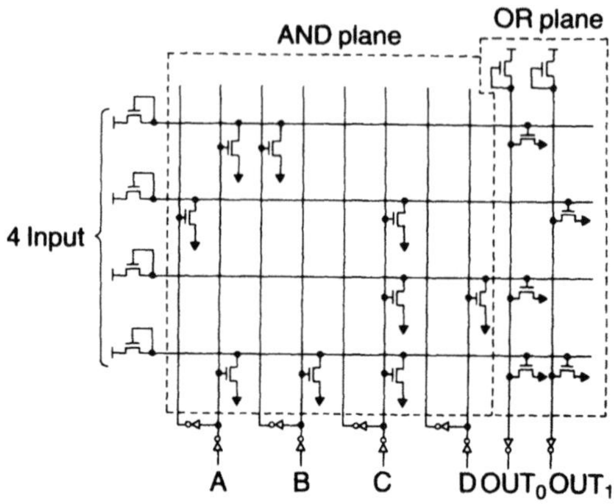

Figure 4.18 An example of PLA.

developed; the first two devices can be used to generate logic arrays having various combinations, and to generate the desired configurations by changing the connection between the lines. Standard cell methods and modular compilers have been developed in order to enhance such devices. This makes it possible to construct VLSI devices (ASIC–Application specific integrated circuit) on the small scale. ASICs are essentially customized semiconductors for a specific purpose and so require special manufacturing procedures.

A PLA has two ROMs, one for the AND and the other for the OR plane. The outputs of the AND are connected with the inputs of the OR which make it possible for a set of AND and OR planes to generate any type of logic function. Figure 4.18 shows an example of logic arithmetic for the expressions:

$$OUT_0 = A \cdot B \cdot C + C \cdot D + A \cdot \bar{B}$$

$$OUT_1 = A \cdot B \cdot C + \bar{A} \cdot C$$

If the ROMs are programmable, they are designated FPLA (field PLA). PAL is a modification of FPLA and was developed by MMI in 1978. The AND plane is programmable but the OR is fixed. PLA has a simpler circuit than FPLA so it has a faster response, as well as simplifying the design. PLA and PAL are sometimes called PLD. Both enhance of the range of application of ASICs.

In gate arrays, the transistors, resistances, diodes and capacitors are employed as standard circuit elements. The function, characteristics and pattern of each standard circuit element are stored in computer memory. When a customer specifies a requirement, an optimal circuit design can be generated from the basic elements. In the first stage, a standard element array is assembled from which a circuit is produced. This provides the detail from which the final metallization stage for wiring can be generated.

In the standard cell method, small scale integrated circuits (SSI) and medium scale integrated circuits (MSI) are provided as standard units which are then combined to give the target functions. Typically this requires 8 to 13 masks (see Chapter 5) for each chip. It takes longer to design a chip than the gate array but is it convenient in developing high quality circuits on a small scale and for small production runs. The advantage of the gate array method is that it uses fewer masks.

The modular compiler is also called a structured cell or structured custom method. It is a compiler for production of optimized VLSI components to consumer specifications. From the specifications of the cells and functional blocks, a software package is used to produce a mask pattern automatically. It is also possible to use it to carry out a simulation in order to examine the performance characteristics. The design strategy is very similar to that of the standard cell but the area is reduced by 35–40%, which makes it very useful for ASICs.

4.3.3 Advantages of VLSI

VLSI makes wire lengths much shorter. In early computers, discrete elements were connected to each other on printed circuit boards. The signal delay time along a wire on a printed circuit is approximately 0.012 ps/μm. The average wire length on a board is 150 000 μm, so the total signal delay is about 1.8 ns. In the case of an IC, the delay time and the wire length are 0.02 ps/μm and 1250 μm respectively, so the total signal delay time is of the order of 0.03 ns. For VLSI, the average delay time can be even shorter, so considerable effort has been devoted to shortening the connections.

4.3.4 Design and assembly of chips

When LSI was introduced, only a few engineers were involved in circuit designs, which were used largely by counters, calculators and desk-top computers based on 4-bit microprocessors. Typically, this required of the order of 13 man-years of effort for 8-bit and 52 man-years for 16-bit microprocessors. It is impossible to design a VLSI system manually and it is necessary to rely on CAD (computer aided design) procedures. Moreover, increased density requires the structure to be changed because new special purpose elements need to be incorporated which means communication between the components becomes important.

CAD requires the functional specification of designs and processing of masks and devices to make use of structured programming. Figure 4.19

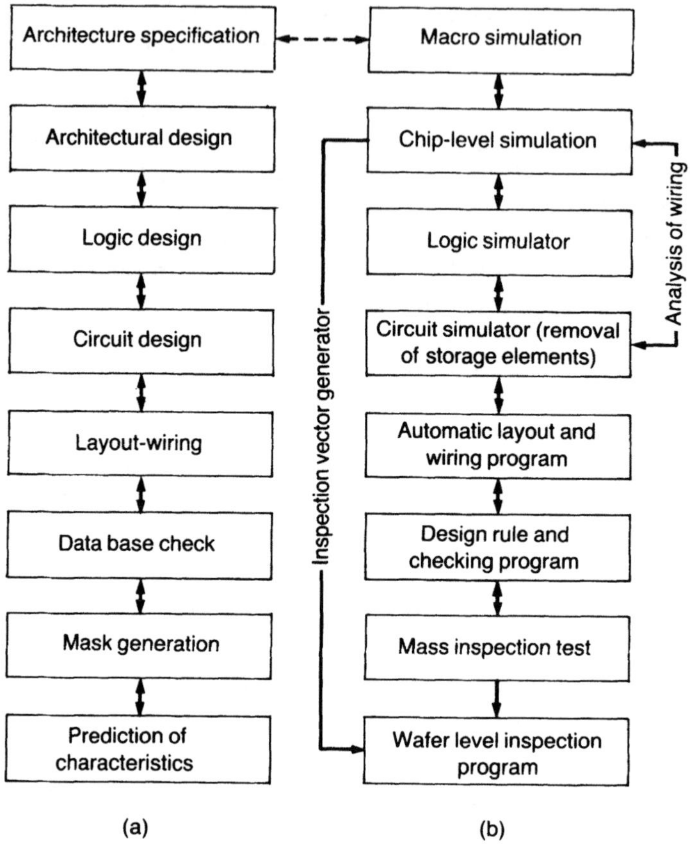

Figure 4.19 VLSI design hierarchy and computer aided design tool: (a) design hierarchy and (b) CAD tools.

shows the correspondence between the VLSI design phases and software tools. The microsimulator evaluates performance and functionality of the proposed architecture and the chip simulator evaluates the models of the hardware: the logic simulator examines the design of the logic gates. A circuit simulator is used to look at the transient response of the current for a block, based on the layout and characteristics such as stray capacitance and the wiring resistance.

4.4 NEW DEVICES

There are two fundamental devices based on VLSI with silicon, namely the bipolar transistor and MOSFET. VLSI needs manufacturing techniques for dealing with devices at the submicron scale and the resulting scaling gives rise to problems which in turn require new devices to overcome them. These issues are discussed below.

4.4.1 Problems of miniaturization

As the line widths of VLSI become smaller, the area of the active elements is reduced and this causes previously unimportant physical phenomena associated with the larger area devices to assume significance, such as the contamination of the signals which cause the performance of the devices to degrade. The effects can be attributed to the following.

1. *Noise*: This arises from the recombination of electrons and holes and cannot be neglected below 0.5 μm. With VLSI, cross-talk arises due to electromagnetic induction and wires must be twisted together to overcome it.
2. *kT/q effect*: Electrons with charge q have energy in proportion to kT, which means it is necessary to maintain a signal voltage several times larger than kT/q (0.0295 V at 300 K). This value can be reduced by decreasing the temperature.
3. *Hot electrons*: A hot electron is one having an energy greater than kT. Hot electrons in a silicon layer invade the oxide-silicon insulator. In the FET, hot electrons change the threshold voltage of the gate and so degrade the characteristics of the FET. This restricts the gate length of the FET to about 0.25 μm. In the case of the bipolar transistor it also causes current leakage at the contact points.
4. *Distribution of impurity concentration*: The atom density of silicon is $5 \times 10^{22}\,\mathrm{cm}^{-3}$ and the upper limit of impurity concentration is about $10^{19}\,\mathrm{cm}^{-3}$. As the area of the device becomes smaller, random distribution of the impurity occurs which can influence the characteristics.
5. *Heat generation*: The power consumption in an element is less in VLSI but because the number of integrated circuits increases exponentially, the total power increases. This necessitates cooling, for example by using a refrigerant gas or an inert liquid.

4.4.2 New developments in devices

To overcome the above problems, especially those arising from short channels, several new devices have been developed.

1. *ALD transistor* [18]: Originally the FET channel was formed on the surface. More recently, it has been formed by implanting a single atomic layer of impurity atoms which can also block the leakage current. Even if the channel is long, it is possible to make the current path width narrow. This eliminates the short channel effect so the FET

Table 4.1 Some devices based on GaAs and superconductors [112]

Name	*Material*	*Structure*	*Feature*
MES FET [23] (metal-semiconductor FET)	nGaAs	Gate, source, drain are metal	4 Kbit SRAM; access time 1.0–1.5 ns 4.8 W
HEMT [24] (high electron mobility-transistor)	AlGaAs/GaAs	Quantum well between hetero-junction of nAlGaAs/iGaAs (FET like type)	Cutoff frequency 350–400 GHz; low noise figure 0.5 dB (18 GHz) for satellite communication
SIS FET [25] (semiconductor insulator semiconductor FET)	GaAs/AlGaAs (undoped)	Hetero-junction is sandwiched between insulator	High speed, high density and suitable for mass-production
HBT [26] (heterojunction bipolar transistor)	GaAs/AlGaAs	Graded junction for emitter-base	High speed, high transconductance 6000 mS/mm; f_{max} = 40 GHz
HET [27] (hot electron transistor	GaAs/AlGaAs	Electron can pass through emitter base potential barrier by tunnelling and reach high kinetic energy.	High speeds are expected; electron can pass base 0.1–0.5 ps
RHET [28] (resonant tunnelling hot electron transistor)	GaAs/AlGaAs	Quantum wells are inserted between emitter and base	Current gain of emitter common is 10–17; f_{max} = 100 GHz; Functional device for parity or A–D convertor

Ballistic transistor [29]	AlGasAs/GaAs	Hetero-junction bipolar transistor structure	f_{max} = 105 GHz; delay time 2.6 ps/gate
Josephson devices [30]	Nb	Superconductors sandwiches oxide insulator	Magnetic control; at He temperature, computing speed and power dissipation are ×10, ×1/1000 of Si computer, respectively
Superconducting transistor [31]	Nb/Si	Superconducting source opposes drain on Si substrate with metal gate is between (FET-like type).	High input impedance. Switching speed 20 ps, power dissipation 5 μW at He temperature
QFP [32] (quantum flux parametron)	Nb/PbIn	2 terminal parametron using Josephson junction pair and Josephson inductor.	At He temperature 100 GHz excitation 1 ps/gate, 1 nW; Various integrated circuit sheets can be coupled without overlapping

with gate lengths of less than 0.5 μm can still meet the required performance.

2. *Three-dimensional devices* [19]: These are formed by building up the basic two dimensional integrated circuit (2-D IC). The 2-D IC is covered with an amorphous insulator and then a new polysilicon layer is formed on it using laser annealing. A third layer is then used to generate a 3-D device. The essential feature of multilayer fabrication is the formation of different circuits on individual layers. This approach is now used in building circuits for use in applications based on artificial intelligence.
3. *XMOS* [20]: This has a sandwich structure having an active layer between two gates. The term is used to indicate the shape of the cross-section. In XMOS, the short channel effect can be reduced by forming a thin active layer. A three-dimensional device made by using this approach has two gates and so prevents electromagnetic interferences between the upper and lower layers.
4. *Systolic array* [21]: This device has been developed in order to process large volumes of data. High speed computing is then possible by arranging many elements in a regular array and then transferring data between them in synchronization.
5. *Bi-CMOS* [22]: Bipolar transistors and CMOS can be combined on the same Si substrate by using standard lithography. Bi-CMOS is used for high speed memory requiring low power consumption (256 Kbit SRAM and 1 Mbit DRAM) in applications such as in video-camera controllers, D/A converters, floppy disk controllers.

Devices based on high speed GaAs or super-conductors are shown in Table 4.1.

5

VLSI process engineering operations

Before giving any description of semiconductor process technology, it is useful to understand what is involved in VLSI manufacture. This can be most easily understood by considering the design of the manufacturing process for CMOS DRAMs and is appropriate because this is the most common semiconductor made in Japan*. In developing the process, the equipment to be used in each elementary process, the mode of operation and time of processing needs to be considered. Also, the necessary complex production control of the manufacturing process needs to be discussed. All of this must be set in the context of the market for silicon wafers and chemicals which have to be supplied by the chemical industry, the influence of dust on productivity and the environmental problems arising from the manufacturing process.

5.1 ASPECTS OF PRODUCTION

5.1.1 Process flow diagram

CMOS 1MBIT-DRAM [33]

The DRAM is a good example of VLSI devices because it represents the highest level of integration and is manufactured on the largest scale. This will probably continue in the immediate future and so it is a good working model for gaining insight into VLSI process technology.

A unit memory cell of the DRAM consists of a transistor and a capacitor. The transistor switches an electric current flowing to (or from) the capacitor on and off. The state of the capacitor depends on the

* This reflects the origin of the book which was directed at chemical engineers working in Japan.

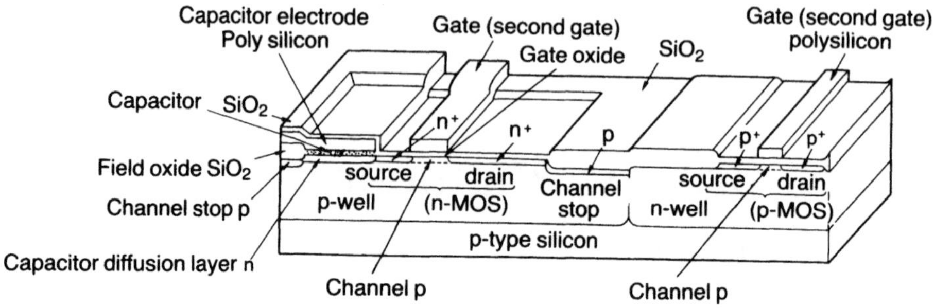

Figure 5.1 Structure of unit memory cell and name of each part: p^+ = positive (high-dose doping of group III element); p = positive (normal-dose doping of group III element); n^+ = negative (high-dose doping of group V element); n = negative (normal-dose doping of group V element).

electric charge which accumulates on the electrodes, and is identified by switching the transistor and monitoring the voltage generated across the capacitor.

The metal oxide semiconductor (MOS) is a three-layered structure which operates as a field effect transistor (FET), the details of which are illustrated in Fig. 5.1. As the name suggests, the operation is dictated by the electrostatic field on either side of a very thin (30 nm) insulating layer (gate) produced in a subsurface channel between the source and the drain. These can be seen as being equivalent to the base, emitter and collector of the conventional bipolar transistor. The current flow between the source and drain is controlled by applying a voltage (normally 5 V) to the gate which allows the device to act as a switch with an operating time basically determined by the path length and the mobility of the electrons or holes (depending on whether it is a p-n-p or n-p-n device). The capacitors are formed by producing an oxide layer about 10 nm thick in a substrate of heavily doped semiconductors. This similarity in structure of the FET and capacitor has the advantage that the manufacturing operations for the MOS DRAM are simpler than those required to fabricate bipolar devices and so make it easier to achieve higher levels of integration.

Complimentary MOS (CMOS) circuits require less electrical power since they combine n-type and p-type transistors which means that one is on while the other is off. This mitigates the effect of Joule heating and so permits a higher degree of integration. 1 Mbit requires the assembling of 10^6 unit memory cells on a chip area of about $0.5\,cm^2$ which implies a unit memory cell can occupy an area as small as $7\,\mu m \times 7\,\mu m$. This makes it possible to have a CMOS 1 Mbit DRAM device consisting of one million cells covering an area $7\,\mu m$ square and electrically connected through pins attached to the chip so that electronic information can be exchanged with

Table 5.1 Time elapsed for wafer processing

Process		*Elapsed time (min)*	*per lot (min)*		*Iteration number*
Cleaning		540	45	×	12
Ion implantation		1 000			7
Oxidation		537.5			8
CVD	Polysilicon	260	130	×	2
	PSG	40			2
	Si_3N_4	440	220	×	2
	p-SiN	90	90	×	1
Al spattering		34	17	×	2
Annealing		355			5
Lithography		2 632.5	202.5	×	13
Coating and Prebaking		22.5	22.5	×	1
Etching	Polysilicon	200	100	×	2
	PSG	137.5			3
	Al	100	50	×	2
	Si_3N_4	249.8			5
	Resist	520	40	×	13
	p-SiN and SiO_2	50	25	×	2
Total		7 208.8			

other circuits. The unit cell of the DRAM which will be used in the case study is shown schematically in Figure 5.1.

Flow diagram and elapsed time

Before looking at any of the operations, it is necessary to examine the flowchart for the complete wafer process and consider the elapsed time for each of the elementary stages, as indicated in Table 5.1. The input to this part of the process follows on from the VLSI design and after growing the single crystals. Figure 5.2 shows the function of each device on the wafer for the various stages along with the elapsed time for that step. The example illustrates the case of a 1 Mbit CMOS DRAM with double-sided wells. This process uses 25 wafers from a 6-inch p-type ingot to form one lot. While the actual industrial process for producing 1 Mbit DRAMs is not exactly the same, it is very similar.

As can be seen from Fig. 5.2, 21 different sub-processes are involved in making a device. However these sub-processes are based on only 7 elementary technologies: cleaning, oxidation, ion implantation, thin film formation (including aluminium), lithography (resist processing) and etching. The formation of the circuit functions on the wafer is by

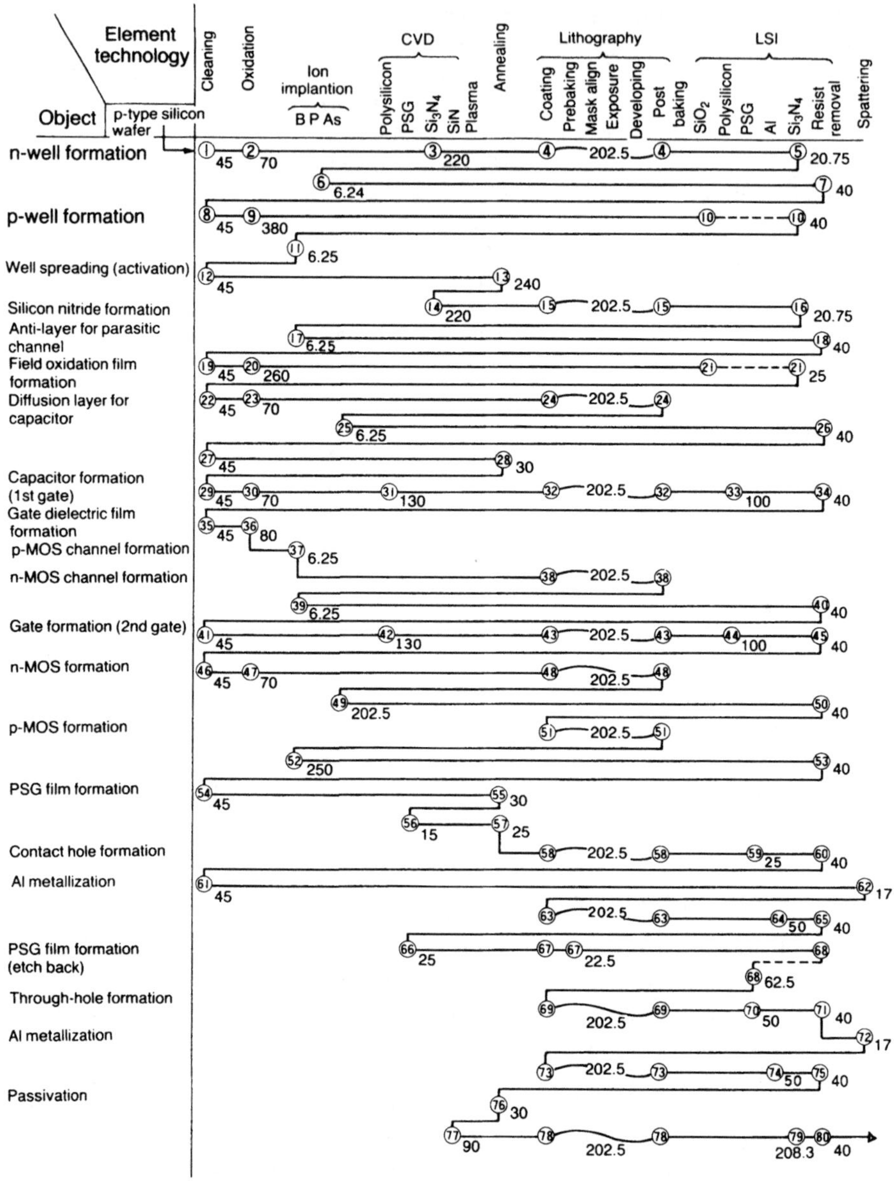

Figure 5.2 Flow diagram of semiconductor process: ○ = process number; subscript numbers = time elapsed for specified process (min); --- = simultaneous treatment; ~ = whole process of lithography.

successive application of the elementary steps. The way in which they are used is strongly influenced by the subsequent operations as well as any factors imposed by the device structure. In addition it is important to take account of anything which may require operating conditions which have a deleterious influence on some part of any device which has already been made. In such cases, the affected parts may need a protective film before any processing takes place. Even though one unit operation is used several times, the actual operating condition will be different in each case.

The manufacturing flow diagram is given in Fig. 5.2 in some detail. The rows and columns correspond to sub-processes and elementary processing stages respectively, with the number of each sub-process denoting the sequence in the circle. These are cross-referenced in the following description by inserting the corresponding numbers in brackets. (See Fig. 5.3 also.)

(a) Formation of a self-aligned well (trench)

Since the surface of a new wafer may be oxidized naturally, the surface layers need to be removed carefully by cleaning and etching. This sequence of operation is called preliminary cleaning (1). The pad oxide film for LOCOS (local oxidation of silicon) is formed by thermal oxidation (2) at about 1000°C.

$$Si + O_2 \rightarrow SiO_2$$

A silicon nitride film (Si_3N_4) of 100 nm is built up on the pad oxide using CVD (chemical vapour deposition) (3) at 750–800°C and 0.1 torr.

$$3SiH_2Cl_2 + 4NH_3 \rightarrow Si_3N_4 + 6HCl + 6H_2$$

The silicon nitride forms a protective layer for the oxide film transistors which is in tension. The pad oxide reduces dislocations and slip in the silicon wafer and so relaxes the stress.

Formation of an n-well (trench)

The lithography process (4) removes the silicon nitride film where the n-well will be formed. This process involves photoresist coating, pre-baking, exposure, developing and baking. After cleaning the wafer with dimethyl-chlorosilan, the surface is coated with a photoresist polymer to a thickness of about 1 μm by feeding it on to a wafer rotating at 4500 rev/min. To remove solvent from the polymer layer, the wafer is heated to 85°C which causes the resist to solidify. A pre-formed mask is attached to a projection aligner and is precisely fixed in the correct position, after which it is exposed and developed. The resist layer which has been exposed to the ultraviolet beam is destroyed, leaving the remainder behind. The residual resist layer is hardened by heating to 135°C

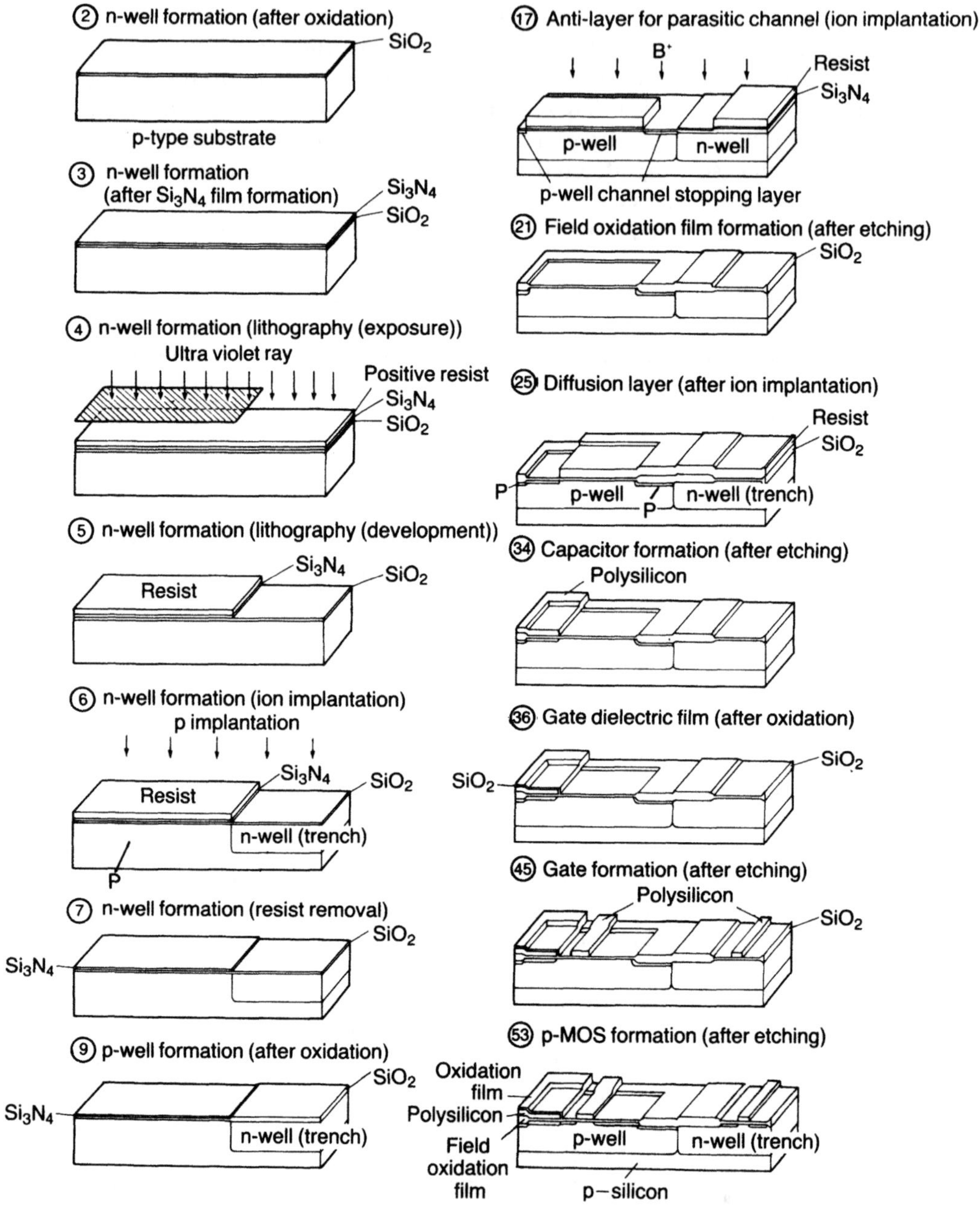

Figure 5.3 Changes to the device made during manufacture. The encircled numbers correspond to numbers in the flow diagram.

(baking). This process is fully automated so as to protect the resist layer from dust contamination.

The wafer then passes to the etching process (5) where the silicon nitride film is removed (at about 120 nm/min) by etching with a plasma of CF_4 and O_2 at a pressure of 1 torr.

$$Si_3N_4 + 12F\cdot \rightarrow 3SiF_4 \uparrow + 2N_2O$$

At these operating conditions, the oxide film is weak and the etching rate is only about a half that of silicon nitride.

The wafer moves on to the implanter where ions accelerated by a potential field in the range 25–200 keV, are embedded in the oxide film to give a p-ion concentration in the n-well of about 10^{12}–$10^{13}\,cm^{-2}$ (6). The depth of implantation can be controlled by the ion acceleration. Any remaining resist is then burned to ash in an oxidizing atmosphere (7).

(a) Formation of a p-well (trench)

After cleaning (8), an oxide film of 150 nm is formed on the area which has not been covered with silicon nitride during thermal oxidation (1000°C) (9). The films of silicon nitride (100 nm) and the upper part of the oxide (about 10 nm) are then peeled off using plasma etching with CF_4 and O_2 (10). The etching process for the silicon nitride film is the same as in stage (5). The reaction scheme involved in forming the oxide film takes place at 1 torr according to:

$$3SiO_2 + 4CF_3\cdot \rightarrow 3SiF_4 \uparrow + 2CO_2 + 2CO$$

The B^+ ion implantation field operates at 25–250 keV which produces a concentration in the p-well of around 10^{12}–$10^{13}\,cm^{-2}$ (0.01–1 mA) (11). This operation disorders the crystal structure of the wafer and the ions are dispersed randomly. After cleaning (12) and annealing at 1200°C, so as to arrange the ions in the lattice, they are diffused horizontally up to 3 μm (13).

(b) Formation of the silicon nitride film and the anti-layer for the parasitic channel

An anti-layer for a parasitic channel is deposited in order to prevent an inversion layer from forming near the surface of the oxide film. Following the procedure used in (3) above, the silicon nitride film is built up (14), and the resist which covers the region where the anti-layer will be formed is removed using lithography (15), followed by etching of the silicon nitride (16). In step (17), the B^+ ions are implanted under the same operating condition as for (11) and any remaining resist is then burned off (18).

(c) Formation of the field oxide film

Following cleaning of the wafer (19), an 800 nm oxide film is formed by introducing steam at 1000°C (20) which gives rise to the following reaction:

$$Si + 2H_2O \rightarrow SiO_2 + 2H_2$$

The silicon nitride film is unaffected but the silicon under the edge of the film is slightly oxidized. Because of the increase in volume of the silicon due to oxidation, the nitride film bends backwards while remaining bonded to the oxide film. This gives rise to a shape called the bird's beak. After this the silicon nitride and oxide films are removed by etching (21), as in stage (10).

(d) Formation of the diffusion layer for a capacitor

After further cleaning the wafer (22), a very clean new oxide film is formed (23). Lithography (24) is used to form a mask and the As^+ ions are implanted at 200 keV. This results in the ion concentration in the diffusion layer of 10^{13}–10^{14} cm^{-2} (0.01–1 mA) (25). The resist is removed (26) and the wafer cleaned again (27), followed by annealing around 1000°C in an atmosphere of nitrogen (28).

(e) Formation of capacitor (the first gate)

Following careful cleaning of the wafer (29), an insulator film of 10–2 nm is formed at 800–900°C using O_2 and steam (wet oxidation method) (30). After this, a polysilicon layer about 400 nm is built up on the insulator film using CVD (31) at 0.5 torr.

$$SiH_4 \rightarrow Si + 2H_2$$

A polysilicon film foundation for the capacitors is deposited using lithography (32) and etching (33). Polysilicon etching is carried out using CF_3Cl and O_2 at a rate of about 100 nm/min, which takes place by the following reaction

$$Si + 4F\cdot \rightarrow SiF_4 \uparrow$$

after which the resist is removed (34).

(f) Formation of gate dielectric film

A very fine oxide film for insulation is formed at 800–900°C using wet oxidation (36).

(g) Doping of n/p-MOS and p-MOS channels

A channel is created between the source and drain, which is then used to form a gate. B^+ ions are implanted under a potential of 25–250 keV (0.01–1 mA) which keeps the ion concentration in the channel within the range 10^{11}–10^{12} cm^{-2} (37). The procedure for doping the p-MOS channel is similar to that used above. The B^+ ions are implanted in the channel (39) and then the resist is removed (40).

(h) Formation of second gate

This gate acts as a switch controlled by an external voltage. After carefully cleaning the wafer, a polysilicon film is formed (42) following a procedure similar to (31). Part of the film to build up the gate is formed by lithography and etching during stages (43) to (45).

(i) Formation of n-MOS

The source and drain (n-MOS) are formed from a p-well. After cleaning, a fine oxide film is formed using the same procedure as (2) so as to prevent channelling, while ensuring the resistance is high (47). Using lithography (48), two windows for the source and the drain are made in the resist. As^+ ions are implanted at 10–150 keV to give an ion concentration in the n-MOS of 10^{15}–$10^{16}\,cm^{-2}$ (10 mA) (49), after which the resists are burned off (50).

(j) Formation of p-MOS

The resist located at the p-MOS for the n-well is removed by lithography (51). B^+ ions are then implanted at 10–150 keV to keep the concentration in the p-MOS to within 10^{15}–$10^{16}\,cm^{-2}$ (about 10 mA) (52). The resist is then burned off (53).

(k) Formation of the PSG (phospho-silicon glass) film

A PSG film is used to completely cover all the parts already made. After cleaning (54) it is annealed at 900–1000°C in a nitrogen atmosphere (55). This operation is necessary for activation of the implanted ions. A PSG film about 500 nm thick is formed by CVD SiH_4PH_3 in O_2 and N_2 (56) at 1 atm., and 400–500°C by the reaction:

$$SiH_4 + 2PH_3 + 6O_2 \rightarrow SiO_2{\cdot}P_2O_5 + 5H_2O$$

The concentration of P needs to be in the range 4–10% and the wafer is annealed at about 950°C in a nitrogen atmosphere (57). The annealing makes the grain size of the polysilicon uniform and flattens the surface of the film, after which the PSG is dehydrated.

(l) Formation of contact holes

The contact holes for the gates and capacitors are made through the PSG film. The resist covering the holes is removed (58) and the PSG etched (500 nm/min) at about 2 torr using CHF_3 and small amounts of CF_4 and He (59).

(m) Aluminium metallization connections

Aluminium is very useful for making connections because it gives a very fine film which adheres to the substrate. In addition, it also offers excellent step coverage. An aluminium strip about 1 μm is formed by sputtering (62) (at a rate of 60 nm/min). The resist corresponding to the connection lines remains after lithography (63) and any aluminium is etched with CCl_4 (or $SiCl_4$, Cl_2 and heat about 500 nm/min) to remove everything but the lines (64).

(n) Formation of PSG film (etch back)

The PSG film is used as an insulating film to make the surface flat. Because the surface is very rugged, it is difficult to make connections. One method is to make the surface flat by etch back. A PSG film 1.5 μm thick is formed at 400–500°C using SiH_4, PH_3, O_2 and N_2 (66). After coating and prebaking, the film is formed using lithography (67) to etch both the resist and PSG (68).

(o) Formation of through-holes (interlayer connection)

The aim of this step is to make holes for connection to the existing links. The resist remains except for the places where the holes are to be formed (69). The PSG film is etched using CFH_3 at about 500 nm/min after which the resist is burned off (71).

(p) Aluminium metallization connections

Using the same aluminium sputtering as (62), a metallic film is formed on the wafer surface (72) and the connections are completed using the same procedures as (73) to (75).

(q) Passivation

This process is the final step in removal of O_2. The wafer is annealed at 400–450°C with H_2 (76). A p-SiN film 1 μm thick is formed as a protective film from SiH_4, NH_3 and N_2 at 250–300°C using plasma CVD. The film prevents the chip from degenerating due to environmental conditions (77). The film is flattened using lithography and etching, after which the remaining resist is removed ((78) to (80)). The wafer is now ready for assembly, the first step of which is cut it into chips one at a time. Each chip is fixed in a container (tie-bonding) and the electrodes of the container are connected to the electrodes (bonding). The whole chip is completely covered with plastic and the identification of the chip and the maker's name are printed on the outside.

5.1.2 Elapsed time

The figures to the right of the circle in the flow diagram show the elapsed time for each process, excluding waiting and inspection. In general, there are two types of processing equipment: single wafer and batch processing. The former treats each wafer individually and is suitable for large diameter wafers. The latter handles one or two lots of wafers. The batch processing equipment can be used for etching, annealing, oxidation, deposition and CVD. As a result of recent developments, the equipment for etching, deposition and CVD tends to combine all these processing stages. The calculation of elapsed times is based on a single wafer unit, though some batch processing is still used. Since both types of equipment are involved, the elapsed time for all the processes needs to be recalculated in terms of lots (1 lot = 25 wafers).

Cleaning takes 45 minutes, but there are several possible methods, each of which has a different elapsed time. Since initial cleaning is longest, this is taken as the elapsed time. In thermal oxidation, the elapsed time for loading, unloading, heating and cooling is about 60 min. Heating has to be carried out very slowly, otherwise the wafer may crack due to thermal stresses. It takes 10 min to make an oxide film of 35 nm, but where a film of 150 nm is necessary, as for the well, this increases to 320 min. This difference is because diffusion is limiting the supply of oxygen to the oxide layer. Wet oxidation takes about 10 min to form a film about 10–12 nm thick.

The formation of a silicon nitride film 100 nm thick takes about 100 min but the heating and cooling of the wafer requires about 120 min. Polysilicon films take 30 min to form a film 400 nm thick and 100 min to adjust the pressure. In the case of PSG, a film 500 nm can be generated in 5 min but it takes 10 min to change the temperature. The elapsed time for etching back a 1.5 μm PSG film is 15 min, while the change in operating temperature takes about 10 min. The passivation films of silicon nitride formed by the plasma CVD require about 40 min to adjust the temperature and pressure. However, little time is required for setting the operating conditions for sputtered aluminium films. To diffuse the low concentration impurities during annealing in well formation requires a long time because of the need to heat up to a high temperature and anneal the wafers.

Lithography involves five sub-processes, but the coating and pre-baking are operated continuously and automatically. It takes 0.5 min and 7.5 min to coat one wafer and one lot, respectively. With pre-baking, it is possible to treat one lot of wafers in 15 min compared with 22.5 min if operated sequentially. The elapsed time for exposure is 5 min per wafer, i.e. 125 min per lot. Developing and baking can be carried out sequentially with times varying from 1 min per wafer to 30 min per lot, respectively. Using a lithography, the time is 202.5 min. Times for etching depend on the materials and operating conditions, as can be seen from the list below.

	Time in min
Silicon nitride	0.83
Silicon nitride and oxide	1
Polysilicon	4

PSG	1
Aluminium	2
Resist and PSG	2.6

The elapsed times for all these stages are summarized in Table 5.1. These values do not include the waiting and inspection times so they are the net times for each process. Table 5.1 suggests the throughput time is 5 days. In practice, it takes about 30–40 days to finish completely one lot of wafers. The difference is mainly because of the time involved in inspection, which generally has to be carried out manually. The time difference is still large and is due to the fact that equipment is often not reliable and so causes delays. Also because of rapid improvement in equipment, there tends to be frequent replacement and it takes a long time to tune the operating conditions. Since the investment cost is very high, a long waiting time is allowed to try to ensure each part of the plant works at maximum efficiency.

5.1.3 Production control

The processes and equipment used in VLSI manufacturing are quite different from those of the chemical industry. Hence, the factors which need to be considered are also different, particularly in respect of process design, operation and management. VLSI manufacturing has many stages but the essential unit processes are relatively simple but used many times. Technical innovation of the equipment is very rapid, but reliability is low so effective utilization is therefore very low. The whole process involves both batch and single wafer processing, so it is difficult to control the total flow of wafers and some arrangement for transfer between each stage is needed, which makes it very difficult to keep a very clean environment.

In terms of production, the important factors should be throughput, total elapsed time and yield. Generally, if the number of wafers in the line decreases, the total elapsed time is shorter. On the other hand, as the number of wafers increases, the processing time is longer although there is a greater throughput. A longer total elapsed time results in the longer waiting time for this stage. With a further increase in the number of wafers, even improving the throughput of the equipment at the bottle-neck does not increase production. The throughput and elapsed time for processing are closely related to the bottle-necks in the production rate (flow rate of wafers). For example, to change the measurement techniques in the production system requires consideration of the number of wafers in the process line, the number of stages, the nature of the processing and the layout.

The yield is very important because it is directly related to the economics. It is influenced by factors such as

- defects in the film
- lack of standard operating conditions
- failure of processing equipment
- training of operators

As the level of integration increases, the density of defects in the film seriously affects the yield, mainly due to dust. These defects make pinholes, cracks and projections which give rise to short circuits and degradation of the film. It is therefore very important to ensure a clean processing environment and to plan for the introduction of automation.

Careful management of process conditions can affect the other three factors. The parameters which need to be controlled are: process variables, device characteristics and design rules. The process parameters specify values such as the depth of connections, sheet resistance, concentration of impurities and so on, while device parameters characterize the device, e.g. threshold and capacity. The design rules specify the width of the line pattern, the line pitch, extension and compilation. To standardize operations, it is necessary to know the values of the process variables as well as understanding the relationship between them. The setpoints for each process are based on calculated values, although the operational values tend to be slightly different. In order to minimize deviations, the operating conditions need to be set in terms of the monitoring devices as well as operating data such as temperature, pressure, concentration and operating time. To prevent mal-operation, computer-based management systems, which automatically set up all the operating conditions, are becoming very important.

5.2 BASIC PROCESSING TECHNOLOGY

5.2.1 Cleaning

Purposes of cleaning

As shown in the case study for the process for the 1 Mbit DRAM, contaminants are generated during plasma etching, resist ashing, annealing and ion implantation followed by deposition on the wafers. Cleaning is therefore necessary to remove them in order to increase the yield of products. This is done using both physical and chemical processes. However, the first step is to try to avoid formation of the impurities in the first place. A prerequisite is a dust-free atmosphere which calls for careful design of clean rooms, avoiding contamination from particles arising from the equipment and during transfer operations. When contamination does occur, it is necessary to know precisely what form it takes in order to determine how to deal with it. In this respect, the problems are similar to those encountered in cleaning chemical plant. However, there are no standard procedures and each cleaning operation must necessarily be individually designed.

The contaminants adhering to the wafers can be classified into ionic and non-ionic. Ionic contaminants, such as alkali metals, adsorb on the surface during acid and water rinses. Non-ionic contaminants can be either organic or inorganic. The organic materials arise from resist residues during processing, from clothing in the clean room, secretions such as dandruff and perspiration from workers during handling of the wafers and dead microorganisms in the chemicals. Inorganic contaminants are mainly heavy metals and oxides produced during the etching process and impurities in the chemicals. The purity of chemicals and rinse water is therefore very important. Any particles produced in the process equipment are physically deposited on the wafers. It is generally considered that workers are the main source of particles so any developments in process automation can be expected to reduce contamination from this source. Periodic cleaning of equipment and implements is a necessary first step and this includes lithography masks, reticules and quartz tubes used in the diffusion process as well as the wafers.

Any dust particle larger than one tenth of the pattern line width is harmful. In the case of 1 Mbit DRAM, this is 0.1 μm. The way in which adhesion of the dust takes place depends on the inertia of the particles for those greater than 1 μm: for particles less than 0.3 μm it occurs by Brownian motion which involves collision between particles and gas molecules. Between 0.3 μm and 1 μm, both processes are involved. Thermal migration and static electricity on the wafers and particles also play a part. When particles reach the surface of the wafer, they are adsorbed physically under van der Waals forces but are made more difficult to remove by any chemical adsorption of moisture from the air.

Wet cleaning

Wafer cleaning methods can be conveniently grouped into wet and dry. Wet cleaning is most commonly used nowadays and the techniques are the same as the chemical and physical operations used for process plants. They can be considered under the following headings.

Chemical methods identified by the cleaning mechanisms

1. *Solvent rinsing*: The ionic contaminants are removed by rinsing with pure water. The removal efficiency increases with increase in water temperature. Heavy metal elements (Cu, Au, Fe, etc.) have little tendency to ionize and so exist as atoms on the silicon surface. They can be ionized by acidic oxidizing reagents, (HCl, H_2O_2, etc.)

The solvents should have a high solubility for the organic contaminants. Typical reagents are hydrocarbon halides such as trichloroethylene (trichlene, $CHCl_2 = CCl_2$), tetrachloroethylene ($CCl_2 = CCl_2$), 1,1,1 trichloroethane (CH_3CCl_3), methylene chloride (CH_2Cl_2), Flon(chloro-fluoro-carbon)R113 (trichlorotrifluoroethane, $CCl_2F = CClHF_2$) as well as alcohols such as methanol, ethanol and isopropanol, acetone, toluene and surface active agents. Trichloroethylene is considered to be carcinogenic and is often found as a contaminant in ground water. FlonR113 is being eliminated because it is considered to cause a decrease in ozone in the stratosphere and so is responsible for the greenhouse effect in the troposphere. While solvents are suitable for the removal of relatively high concentration of organic contaminants they cannot remove monomolecular layers adsorbed on surfaces.

2. *Dissolution by oxidization*: Oxidization of organic substances followed by dissolving in solution enables contaminants such as CO_2, CO and H_2O to be removed. This method is suitable for the light contamination arising from adsorbed layers, since it is possible to remove them completely in a short time. Oxidizing reagents such as H_2SO_4/H_2O_2, NH_4OH/H_2O_2, and fuming nitric acid are used at elevated temperatures.
3. *Surface layer etching*: Some metal contaminants are dissolved by etching the silicon surfaces with caustic detergents. Choline $((CH_3)3N(C_2H_4OH)OH)H_2O_2$, NH_4OH/H_2O_2 and choline surfactants have been used as detergents. Natural oxide films are easily formed on silicon using oxidizing agents or by immersion in water. They can be removed by dilute hydrofluoric acid or by choline and surfactants.

Rinsing

After chemical cleaning, rinsing with pure water is necessary to remove the detergents used, and this requires high quality pure water. About $1.2\,m^3$ of ultra-pure water is necessary to wash a wafer so that there is a need for efficient processes for producing the water as well as minimizing its usage. This must be achieved without compromising the effectiveness of the cleaning process. Some of the rinsing methods are shown in Fig. 5.4 [108]. The overflow type is most commonly used since it removes particles in the air and from wafers and carries them away in the discharge. This operation takes a long time and uses a large amount of water. It can be improved by using 'rapid dumping', where the water in the vessel is periodically removed and then recycled. Rapid dumping with overflow combines the advantages of both methods and is used in cascade.

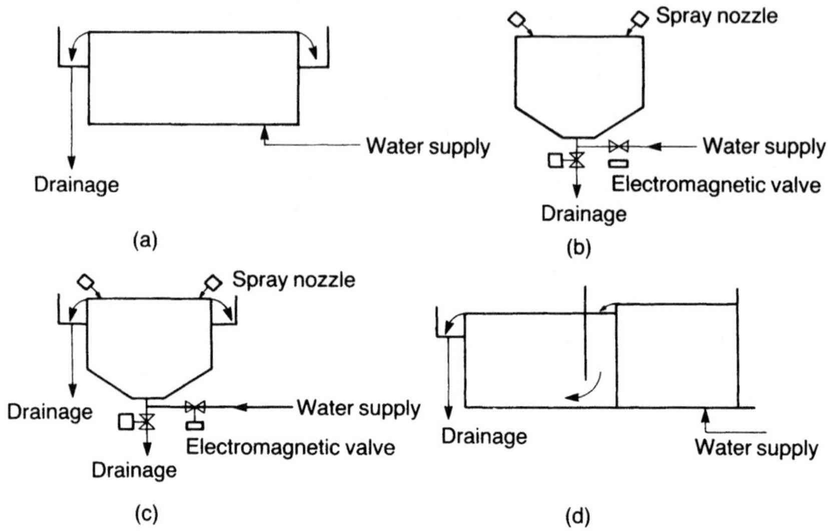

Figure 5.4 Rinsing methods for pure water: (a) overflow; (b) quick-dumping; (c) quick-dumping with overflow; and (d) cascade overflow.

Physical methods: These depend on the method of removal of the particles

1. *Brush cleaning*: A rotating cylindrical brush is held against the surface of the wafer and the particles removed by jetting with detergent solution. The hydrophilic surfaces of the brush hairs are coated by a liquid film so the hydrophilic surface of the wafer is not damaged because it is protected by the presence of the film. This method is suitable for flat, smooth surfaces such as wafers or masks, but not for patterned wafers.

   High pressure water jet cleaning: Adsorbed particles are moved by water jets under high pressure (110–150 $kg\,cm^{-2}$). This method is used in conjunction with brush cleaning, and is suitable for hydrophilic surfaces and patterned wafers. However, because pure water has a high specific resistance, the wafer becomes charged so the specific resistance has to be reduced with dissolved CO_2.
2. *Ultrasonic cleaning*: Ultrasonic waves in the liquid cause a shock wave to be generated during each cycle which first compresses then reduces the pressure of any micro-bubbles. The input energy causes local agitation of the liquid which impacts on the particles. With frequencies of 20–50 kHz, cavitation tends to occur. In the range 800 kHz–1 MHz, there is no cavitation so it is suitable for submicron particles because the vibrating particles enables liquid molecules to get between the

wafer and particle and detach the particle which is then prevented from being re-adsorbed because a liquid film is formed.

Drying

This is the final stage and is most commonly based on spin drying which removes the water from the wafer by centrifugal force. The demand of ever higher degrees of cleanliness is favouring easily vaporized solvents in order to overcome problems from mist generated by the break-up of drops during centrifugal drying. FlonR114 with alcohol or isopropanol has been used but the approach now preferred is to automate the process using hot air drying.

While wet cleaning is suitable for most types of contamination, dry methods are considered better because they avoid problems arising from contaminants in the liquids.

Dry Cleaning. The methods now in use are: oxygen plasma, and ultraviolet rays and ozone

1. *Oxygen plasma*: Oxygen is introduced into a high frequency electric field which results in the organic contaminants being oxidized by the radicals and ozone formed in the plasma. The products are removed as CO_2, CO and H_2O. This method is attractive for the removal of photoresist films but the disadvantage is that the heating elements tend to be damaged by the plasma.
2. *Ultraviolet rays and ozone*: Irradiation with ultraviolet light from a low pressure mercury lamp in the range 185 nm and 254 nm causes organic contaminants to be decomposed or oxidized by the oxygen radicals and ozone which are formed. Much higher cleaning rates are possible compared with the oxygen plasma method because of modern mercury lamps are now very efficient.

Wafer cleaning

The overall wafer cleaning process is a combination of the operations described above. An example of wet cleaning is shown in Table 5.2 [109]. The organic contaminants are removed first, followed by the oxide film. Metal contaminants and ionic alkaline metals are removed by acid cleaning. Particles can then be removed by rinsing with pure water after which the oxide films are treated with dilute hydrofluoric acid and then rinsed and dried. Detergent liquids such as $NH_4OH/H_2O_2/H_2O$ are used for the organic contaminants and $HCl/H_2O_2/H_2O$ for metals. Fundamental research into the mechanistics of the processes involved has been carried out by Kern *et al.* at RCA Laboratories [110] using tracer methods.

As already noted, there are no general standard methods. The

Table 5.2 Cleaning sequence: removal of organic substances (oil, wax, fingerprints, resist)

1	Treatment with organic chemicals
2	Treatment with surfactants
3	Brush water rinse
4	Acid treatment (oxidizing acid)
5	Alkaline treatment
6	Removal of oxide film (removal of residual and natural oxide films etc.)
7	Treatment with hydrofluoric acid
8	Removal of metal impurities
9	Acid treatment
	Removal of ionic impurities
	Removal of particulate impurities (undissolved materials)
10	Ultrasonic and brush rinsing
11	Removal of oxide film, i.e. natural oxide film formed during cleaning
12	Water rinse: drying

procedures are basically dependent on know-how for selection of the composition, temperature and treatment times prior to oxidation, in the way the diffusion and CVD processes are carried out, and in all the subsequent processing. The exception is initial cleaning used for mirror-finished wafer surfaces which is prescribed by ASTM (American Society for Testing and Materials) rule 4 [111] and outlined in Table 5.3.

Evaluating the effectiveness of cleaning

It is rather difficult to evaluate the effectiveness of the cleaning process. The wetting conditions of the surface are sometimes inspected visually to check for particles on the glassy wafer but such procedures are not very efficient. Surface inspection using laser beam scattering is now widely used. However, there is no good method for determining submicron particles on patterned wafers. To assist in the control of the operation rinse water, acids and alkalis are monitored by electric conductivity meters.

Identifying the sources of contamination and the way in which adhesion occurs, relies on Auger electron spectroscopy, ESCA (electron spectroscopy for chemical analysis), X-ray microanalysis, ion microanalysis to examine the films and metal particles. Gas chromatography and microphotographic fluorescent analysis are generally used to detect the organic contaminants.

In future, it will be necessary to develop non-destructive and non-contacting methods in order to assess how effective the cleaning operation has been, especially after the formation of patterns on the wafer

Table 5.3 Initial cleaning of wafer (ASTM rule 4)

1 Place the specimen and other slices in a carrier and immerse in 1,1,2trichloroethylene in a covered quartz beaker housed in a clean hood. Transfer the beaker and contents to the ultrasonic cleaning apparatus for a cleaning cycle lasting 10 ± 2 min at 75°C

2 Remove the carrier from the beaker and load the slices into the wafer scrubber. Operate the scrubber for a while under conditions customarily used in the laboratory performing this task.

3 Remove the wafers from the solvent and inspect them under normal room illumination. If traces of organic residues are present, repeat 1.

4 Immerse the carrier and wafer slices carefully in boiling H_2O_2 in a quartz beaker, for 5 ± 0.5 min.

5 Rinse the carrier and wafers with ten separate volumes of water at 90°C.

6 Immerse the carrier and wafers in HF in a TFE fluorocarbon beaker for 5 ± 0.5 min at room temperature.

7 Rinse again in accordance with 5.

8 Examine the polished surface in accordance with Method F22 to determine if it is hydrophobic.
 8.1 If the surface is not found to be hydrophobic, swab the wafer with detergent solution and repeat the cleaning procedure, beginning from 1.
 8.2 If the surface is found to be hydrophobic, inspect it under a narrow-beam light source in accordance with Practice F523 to determine whether there is any visually detectable particulate matter.
 8.2.1 If no visually detectable particular matter is found to be present on the surface, it can be considered to be clean.
 8.2.2 If any particulate material is found on the wafer surface, swab it with detergent solution and repeat the cleaning procedure from 1.

because this is important in relation to trench construction. As the scale is reduced further, it will assume even greater significance.

Cleaning equipment

Although the RCA method [110] can be regarded as basically well-founded it does not define a standard cleaning procedure or stipulate a liquid composition: it is still necessary to rely on know-how. This has meant that every equipment manufacturer claims some distinctive features such as facilities for adjusting set-up timing, automatic transfer of wafers and control of solution feed rate. For cleaning prior to oxidation and after resist processing, the wafers are handled in batches. In order to design cleaning equipment properly, however, it is necessary to have information about the reactions, diffusion processes and the dissolution mechanisms as well as the knowledge of the interfacial processes involved in wetting of the surface and the way this interacts with the local fluid mechanics.

5.2.2 Oxidation

Purpose of oxidation

At high temperatures and in the presence of oxides, silicon forms a uniform oxidized layer of SiO_2. This layer is physically very stable, and used for a variety of purposes in silicon devices. In the case study, for example, films are used:

- as a masking layer to protect the silicon wafer when adding ions, e.g. for underlaid oxidized layering (2), a capacitor layer in (23) and for ion interception in (9);
- as insulation to separate elements in the form of a relatively thick (800 nm) field oxidized layer in (2), to separate transistors;
- as capacitor insulation – a thin layer (10–12 nm) can be used because of the high dielectric constant;
- as gate insulation in (2) and (36) – a slightly thicker layer is needed than for capacitors because it must also offer a high voltage resistance; and
- as a passive surface to prevent channelling, by increasing the voltage breakdown resistance (47).

The problem is to form a thin layer of oxide over a large area, having a minimum number of faults and good voltage resistance. This is necessary to form the fine patterns needed in VLSI and to be able to work with

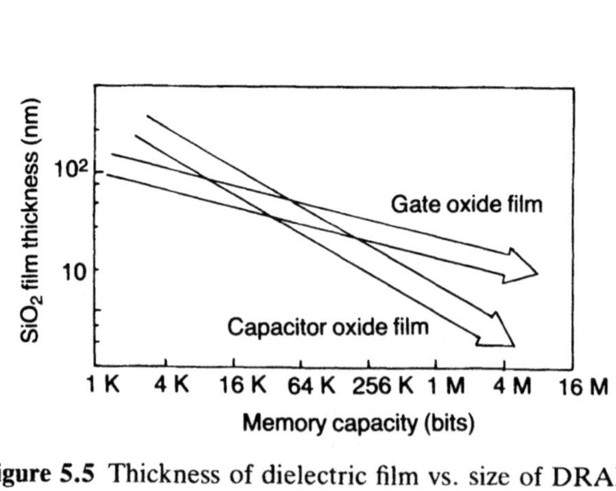

Figure 5.5 Thickness of dielectric film vs. size of DRAM.

large diameter wafers. Figure 5.5 shows how the thickness of the capacitor and gate insulating layers change with the scale of integration [34]. For layers less than 10 nm, the voltage resistance should be capable of being able to maintain an electric field of between 1 and 10 mV/cm. Capacitors now have the thinnest layers because the protection limit in the DRAM is determined by external radiation [35]. To ensure a minimum capacity of 40×10^{-15} F with a 5 V supply, the thickness of the oxide layer needs to be reduced.

Methods of oxidation

A number of models of the oxidation mechanism have been proposed. One regards it as consisting of three phases: gas, SiO_2 and Si with the oxidizing species (O_2 or H_2O) diffusing through the SiO_2 to react at the Si-SiO_2 interface [36]. Such a model accounts for oxidation in atmospheres containing either oxygen or water vapour at temperatures in the range 700–1300°C, and partial pressures of the oxidizing species 0.2–1.0 atm. It produces films of 300–20 000 Å. However, it cannot explain the formation of thin layers. To explain this it has been suggested that ionized oxygen migrates to the Si–SiO_2 interface under the influence of the electric field in the SiO_2 layer [37]. Developments in VLSI technology will require the ability to form very thin high quality oxidized layers of less than 300 Å uniformly and reproducibly. At present, there is no satisfactory explanation which is capable of predicting the relationship between the growth of the oxide layer and the characteristics of the film.

Table 5.4 shows the method for oxidizing silicon used at present. In the

Table 5.4 Main oxidation methods of silicon

Oxidation method	*Oxide reaction system*	*Comment*
1. dry O_2	O_2	Regular oxidation
2. wet O_2	H_2O/O_2	Regular oxidation
3. steam	100% H_2O or H_2O/carrier gas	Regular oxidation
4. hydrogen combustion	H_2O/O_2 ($\rightarrow H_2O$)	Regular oxidation
5. high voltage	H_2/O_2 or O_2	Low temperature, high speed high voltage vessel or reaction in quartz vessel (10 Torr) field oxidation film, isolation
6. plasma	(0) in plasma, radical	from 0.1–1 Torr, [illegible] normal temperature

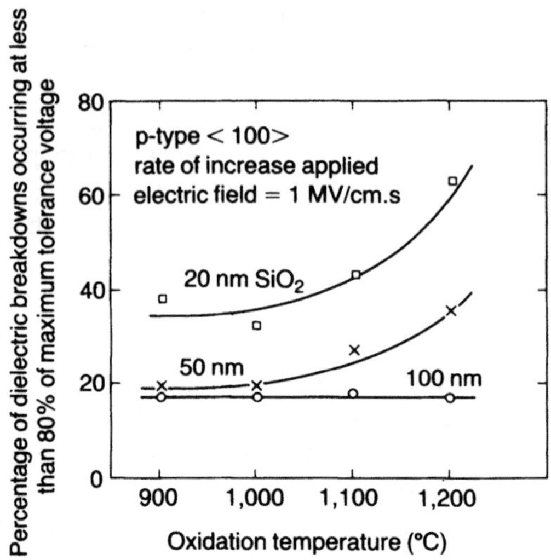

Figure 5.6 Influence of oxidation temperature on the frequency of dielectric breakdown below the tolerance voltage.

case study, it is used for wells in (2), and (9), capacitors in (23) and (30), channelling protection in (11), and for dry oxidation with O_2. Thick layer field oxidation uses oxidation because of the high reaction rates. Gate oxidation was originally based on direct oxidation with O_2 but now uses wet oxidation from H_2 combustion because it can be done at lower temperatures which means there are fewer faults in the film.

The present trend is to oxidize at lower temperatures because this gives films having greater strength, as can be seen in Fig. 5.6 [38] which shows that electrical breakdown increases rapidly with the oxidation temperature. At low temperatures, the reaction rate is slow so the oxygen pressure needs to be increased to overcome this. Compared with other oxidation methods, high pressure oxygen has the following advantages:

- the time to form the film is shorter;
- there is little redistribution of the impurities in the wafer at low temperatures;
- at the Si-SiO_2 interface there are no stacking faults from oxygen insertion in the silicon; and
- the basic physical properties of the oxide layer are uniform.

Plasma oxidation is also carried out at low temperatures but the mechanistic details are still not understood and it is therefore in need of active research.

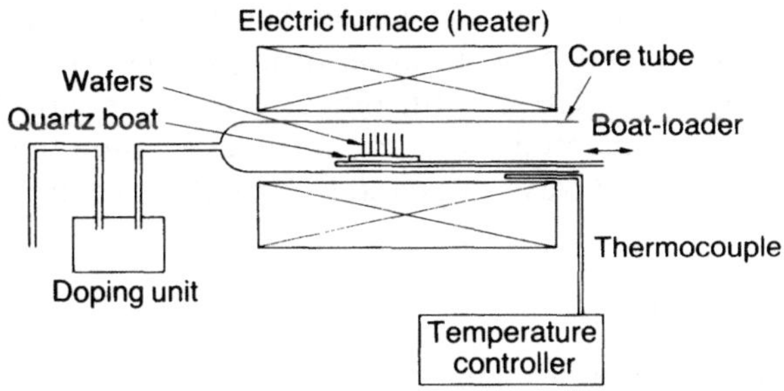

Figure 5.7 Thermal oxidation – structure of diffusion furnace.

Equipment and operation

Figure 5.7 shows an outline of typical equipment used in thermal oxidation. The factors which need to be taken into account are as follows:

1. *Oxidation conditions*: gas composition, atmosphere (wet, dry), amount of water, temperature, time and pressure.
2. *Handling and housing of the device*: the factors of importance are a uniform heating length, the type of wafer insertion board, geometry of reactor tube. For the type of device used in case study, it takes about 60 min to heat and cool when introducing and removing the wafers and any dust must be eliminated.
3. *Wafer processing conditions*: the kind and density of the impurities and the form of the surface.

The strength of the SiO_2 layer depends on all of: temperature during oxidation; the uniformity of conditions and purity during device handling; density of the oxygen layer (determined by Si crystallization) [39]; and particularly the dust formed on the wafer during processing.

The need is to have a sufficiently high rate of oxidation at low temperature with no particles deposited on the surface. It should also be possible to control the temperature so as to have isothermal conditions over an area sufficient to handle the large diameter wafers now being produced.

5.2.3 Ion implantation and annealing

Diffusional implants of group III elements into a silicon wafer are used to change the surface resistance. The impurity produces an electric potential difference between the implanted area and the rest which makes it possible to build a variety of devices. Originally, thermal diffusion was used but this has recently been superseded because it is necessary to have more precise control over concentration of impurities and this is difficult using this technique. These difficulties can be overcome by ion implantation. The implanted ions are introduced into the wafer in a controlled manner and diffuse to give the desired concentration profile.

Aims of ion implantation

As can be seen from the flow diagram of Fig. 5.2 this process is used to make a variety of devices by controlling the number of ions and the depth of penetration. Some examples are given below.

1. *Formation of trenches*: MOS has two types of channel (n and p) and the substrate needs to be charged in order to mark the position where a trench 3–5 mm deep in which channels will be formed and along which the electrons or holes will move.
2. *Formation of an anti-parasitic channel layer*: the electric potential in the line connecting the elements in the oxide layer is influenced by the charges in the oxide film. In some cases this can result in line shorting as the insulation breaks down. Consequently, it is necessary to introduce an interlayer insulator.
3. *Formation of a capacitor diffusion layer* by implanting molecules of As in the capacitor diffusion layer in order to control the resistance.
4. *Channel doping*: the gate voltage threshold is proportional to the number of ions, so it possible to control the threshold by implanting the appropriate number.
5. *Source and drain*: these are impurity layers which provide charges to carry the electric signal. For example, p-type donor impurities such as As can be implanted to give two n-type layers less than 0.3 μm thick.

Principles of ion implantation and equipment

A schematic diagram of ion implantation equipment is shown in Fig. 5.8. The desired impurity molecules are selectively extracted from the plasma in the ion chamber and accelerated in an electric field to an energy level for implantation. It is important to ensure that the temperature of the photoresist used to mask the surface does not exceed 100–150°C. This is done by mounting the wafer on a high speed rotating disk.

Ion implantation is capable of producing a narrow, well-defined diffusion zone reliably and consistently. The depth of penetration is determined by the acceleration energy and the number of ions which is indicated by the integrated value of the beam current. Figure 5.9 shows a typical plot of the relation between the number of implanted ions and acceleration voltage.

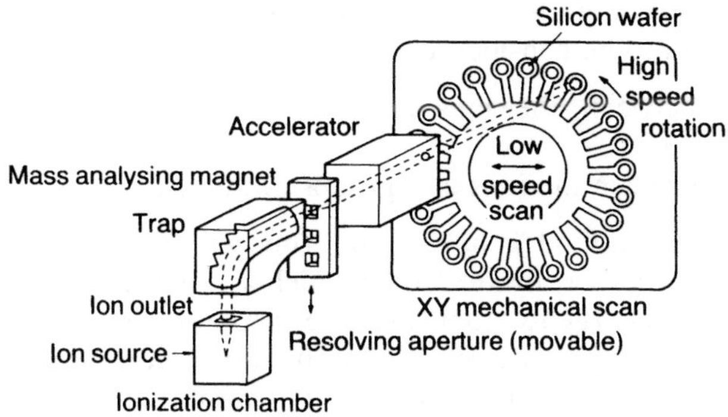

Figure 5.8 Schematic diagram of ion implantation.

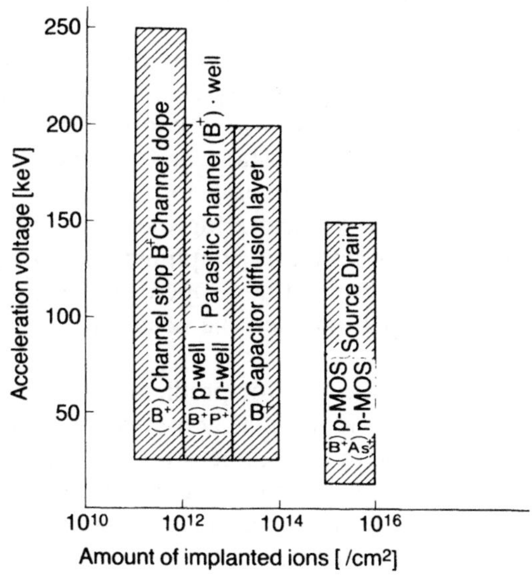

Figure 5.9 Amount of implanted ion and acceleration voltage.

Influence of implanted ions on the substrate

The accelerated ions collide with the atoms on the surface of the substrate and release energy as a result of

- ions stopped by the substrate,
- reflection of ions from the surface,
- collisions giving rise to electron beams,
- collisions causing displacement of atoms at the surface, and
- sputtering of the atoms from the surface.

Reflection accounts for a large fraction of the total. Of the rest, the following need to be considered:

- collision between the ions and single atoms (spherical dispersion model); and
- collision between the ions and groups of atoms.

Several collision models have been proposed. The spherical dispersion model is a two-body collision model which assumes that high energy ions do not transfer energy to atoms because the interaction time is very short so dispersion can be neglected. As the ion enters the substrate, the velocity is reduced and more energy is transferred. The interaction is governed by the interatomic potential, of which there are several currently in use [40].

In the case where the inter-atomic spacing in the wafer is large relative to the size of the ion, it will pass through the solid with very few collisions. One way of modelling this is to regard it as an ion passing through a viscous liquid with the electron cloud of the lattice atoms representing the liquid. The probability is that there will be a collision with the atomic nucleus rather than with the ion [41]. Generally, the rates of both processes tend to be constant and independent of the ionic species. However, with the high energy ions, a collision influences the electron cloud, for example by scattering electrons. In such cases, the rate is no longer constant and this simplified model cannot be used. Lindhard, Scharff and Schiott [42] developed the (so-called) LSS theory based on these models which enabled the positions of the implanted ions in the solid to be estimated.

There are several methods used for predicting the distribution of ions in the substrate [42–47]. Generally, non-crystalline materials have a Gaussian distribution, while in crystalline solids the ions can pass between the lattice atoms which gives rise to two peaks in the ion concentration profile. Channelling through the structure can be avoided by directing the ion beam at an angle to the surface.

With reduction in the size of the cell, the high ion concentration diffusion layer becomes thinner. To form the layer accurately it is necessary to overcome the following problems:

- increase in temperature of the wafer;
- formation of dislocations; and
- channelling.

Annealing

Implantation results in atoms in the wafer being dislodged from the crystal lattice to form a non-crystalline layer. The ions diffuse into the structure until they can find a suitable position in the lattice, after which they are activated by annealing. This is done using an electric heater [48, 49]. The objective of the diffusion process is to achieve the desired depth of penetration to give the desired electrical resistance and concentration profile of impurities in the layer. The annealing process tends to rearrange the crystal structure with the atoms displaced during implantation tending to be concentrated at the phase boundary between the crystalline and non-crystalline layers. After prolonged annealing these atoms are absorbed into the crystal lattice and the excess atoms at the surface of the wafer are released, which gives a stable structure.

If it is assumed that simple diffusion takes place in the crystal, the flux can be expressed by Fick's first law. For a constant diffusion coefficient, with no electric field effects, the concentration profile can be described by Fick's second law. In general, for a given level of impurities, the overall diffusion depends on the concentrations of all the species and the electric field so that the second law needs to be changed to a two-dimensional description which includes the profile parallel as well as normal to the surface. Simulation programs which have been developed to handle these problems generally tend to be unsatisfactory and there is still considerable scope for improvement.

Annealing equipment

The relatively long time needed to anneal the wafers means that it is necessary to treat as many as possible at one time. Typically, about 50 wafers are processed in a 1 m preheating section and then moved to a reaction section for final annealing in an electric oven. The problems which need to be addressed are the loading and unloading of the wafers, ensuring that the flow results in uniform heat transfer, a clean environment and precise control of the operating conditions.

High production rates depend on:

- reducing the time for loading and unloading of wafers as well as minimizing the preheating and cooling down times;
- bringing the reactor tube to a uniform temperature as quickly as possible, which could be as high as 1000°C, depending on the previous stages; and
- establishing a well-defined temperature profile over the whole of the reactor tube.

The rate at which it is possible to increase the wafer temperature is governed by the thermal stresses. As the chips become smaller this needs to be set against the need to keep the thickness of the diffusion layer as thin as possible. Annealing tends to cause re-diffusion of the impurities leading to broadening of the layer, so the annealing time should be as short as possible.

5.2.4 Thin film preparation

Type and role of thin films (Table 5.5)

The manufacturing process involves building up a series of thin films on the surface in particular patterns. The technique used to do this is lithography. Dielectric, semiconductive and metallic films are formed in thicknesses in the submicron range. The thinnest films in the case study are 10 nm for the capacitor insulation layer and 35 nm for the MOSFET transistor formed by thermal oxidation (described in section 5.2.2.). In the following sections, the type, kind, role and preparation methods of the thin films used in the case study will be considered. Descriptions of some other important thin films used in VLSI technology are also included.

Silicon nitride

This is a material containing mainly silicon and nitrogen. These may be in the form of either crystalline Si_3N_4 or amorphous compounds having a variety of chemical compositions. A crystalline Si_3N_4 film is formed by chemical vapour deposition at temperatures above 1200°C and is used as a ceramic coating. The temperature constraints encountered in VLSI processes mean that the films are in the amorphous form. The stoichiometry of films prepared by the chemical vapour deposition at around 750°C is close to that of crystalline Si_3N_4. However, the passivation films obtained by plasma-CVD at 300°C contain much less nitrogen and significant amounts of hydrogen. The amorphous silicon nitride, with stoichiometry close to Si_3N_4 is used as a mask in ion implantation or field oxidation, since the higher density allows it to work efficiently as a barrier layer. The mask is then removed by etching afterwards.

Low temperature processing is essential in the final passivation in order to prevent the stable underlying structure from thermally degrading. With plasma-CVD, an amorphous film of composition SiN_xH_y of lower density is formed by thermal deposition. It is used to protect the underlying structure from moisture and atmospheric contaminants.

Table 5.6 summarizes the properties of the various silicon nitride films and indicates the corresponding process conditions used in their formation.

Polycrystalline silicon

The abbreviated term, polysilicon, is more commonly used for this material. It acts as a gate electrode material for MOS transistors and capacitors. It is formed by CVD from SiH_4 or SiH_2Cl_2 as a single crystal-

Table 5.5 Thin films used in VLSI

Passivation		$Ni_3N_4 \xleftarrow[\sim 0.5\ Torr]{\sim 750°C} SiH_2Cl_2 + NH_3$
		$SiO_2 \xleftarrow[760\ Torr]{\sim 400°C} SiH_4 + O_2$
		$PSG \xleftarrow[760\ Torr]{\sim 400°C} SiH_4 + PH_3 + O_2$
Interlayer insulation	Insulator	$TEOS\ SiO_2 \xleftarrow[\sim 1\ Torr]{\sim 750°C} Si(OC_2H_5)_4$
		$p\text{-}SiN \xleftarrow[\sim 0.5\ Torr]{\sim 300°C} SiH_4 + NH_3$
Gate insulator		$p\text{-}SiO \xleftarrow[\sim 1\ Torr]{\sim 350°C} SiH_4 + N_2O$
Capacitance insulation		Bias-sputtered polyimide
Metallization	Metal	Al, Si, Al, Cu, Si, W
		$WSi_2 \xleftarrow[0.3\ Torr]{350 \sim 450°C} WF_6 + SiH_4$
	Metal	$MoSi_2 \xleftarrow[\sim 1\ Torr]{350 \sim 450°C} MoF_6 + SiH_4$
Gate electrode		$W \xleftarrow[\sim 1\ Torr]{300 \sim 600°C} WF_6 + H_2$
	Semiconductor	$poly\ Si \xleftarrow[\sim 0.5\ Torr]{\sim 600°C} SiH_4$
p-n junction surface upgrading	Semiconductor	$epi\ Si \xleftarrow{1050°C\ or\ higher} SiH_2Cl_2$

line layer which grows epitaxially at temperatures above 1000°C. At lower temperatures in the region of 700°C, polycrystalline films are produced. These films act as an electrically conductive material even though silicon is a semiconductor. This is made possible by doping the polysilicon with a high concentration of an impurity such as boron.

Table 5.6 Silicon nitrides deposited by plasma CVD and LPCVD

	Plasma	*LP CVD*
Substrate temperature [C°]	200 ~ 350	700 ~ 900
Composition SiN_xH_y	x = 0.8 ~ 1.2, y = 0.2 ~ 0.25	x = 4/3, y = 0.04 ~ 0.08
Density [g/cm^3]	2.4 ~ 2.8	2.8 ~ 3.1
Package voltage [V/cm]	$2 \times 10^5 \sim 6 \times 10^6$	10^7
Etching rate [A/min]		
49% HF, 23°C	1500 ~ 3000	80
85% H_3PO_4, 180°C	600 ~ 1000	120

Etching rate is often used in the VLSI field as a measure of hardness, density, chemical stability etc.

Aluminium

This is currently the most commonly used material for metallization (circuit connection). It is modified chemically by partial silicidation ($AlSi_x$) and then slightly alloyed with copper to improve resistance to oxidation, thermal migration and mechanical stress. A disadvantage of aluminium is that it tends to migrate in strong electric fields (electro-migration). With the present 5 V sources used for transistors and capacitors, 5 volts applied over 0.1 micron gives rise to a field strength 500 000 volt cm^{-1}, which gives rise to electro-migration.

Aluminium has good electrical conductivity but poor physicochemical properties. Alloying can partially compensate for this but only at the expense of conductivity. Essentially, the problem originates from the weak chemical bonding of the atoms because the strength is related to the melting point. Processes to develop refractory metal films from W and Mo are now being investigated which will require, among other things, silicidation of the refractory metals to ensure adequate resistance to oxidation.

PSG, BPSG

Amorphous silicon oxides mixed with P (PSG) and P and B (BPSG) give glassy films which are used for interlayer insulation. The reason for introducing P and/or B is to lower the melting point. Surface wafers have a high aspect ratio so the low melting temperatures of PGS and BPSG make it possible to flatten the films thermally. As can be seen in Fig. 5.10, the film can be smoothed using post annealing by a process known as 'glass flow' which is carried out at 950°C for PSG at 900°C for BPSG.

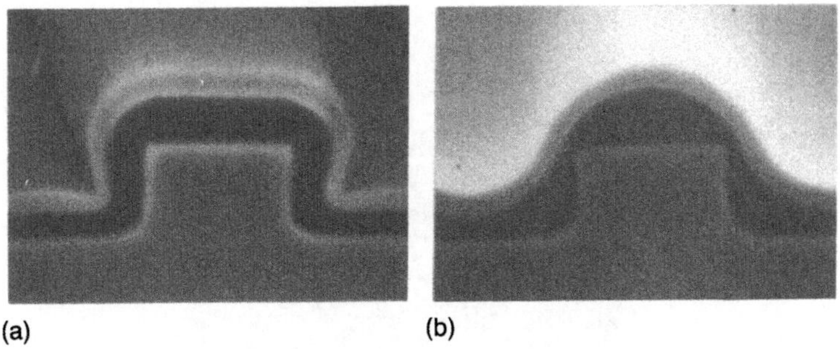

Figure 5.10 Flattening by glass flow: (a) before glass flow; (b) after glass flow.

Epitaxial silicon

This is an important material, although it is not used in the DRAM process. It involves growing a single crystalline Si layer epitaxially on a silicon wafer at approximately 1100°C (c.f. 750°C used in preparing poly-Si). The major areas of applications are:

- construction of the p-n junctions used extensively in the bipolar devices; and
- removal of the crystallographic faults in silicon wafers, which are formed during fabrication from the single-crystal rod.

It is worth noting that most of the thin films prepared in the VLSI technology are either metal or insulator, in spite of VLSI being concerned with semiconductor devices. The surface sublayer of the silicon wafer is used as a semiconducting layer of a transistor. Epitaxial Si is an exception in the sense that a semiconducting active layer is formed during preparation of the film (Fig. 5.11).

Mechanisms of film preparation processes

Almost all films are formed from the gas phase except for organics such as polymides and the sol-gel glassy films prepared by liquid-phase polymerization. Although many sophisticated processes have been proposed and studied only plasma-CVD and sputtering methods have been extensively adopted.

The fundamental mechanisms involved in film deposition are thermochemical reaction and plasmachemical reaction of the gaseous reactants and sputtering of atoms from a solid surface.

CVD

Figure 5.12 shows the chemical reactions occurring in CVD process of Si. A gas such as $SiCl_4$ is used to grow a film in a special reactor where the

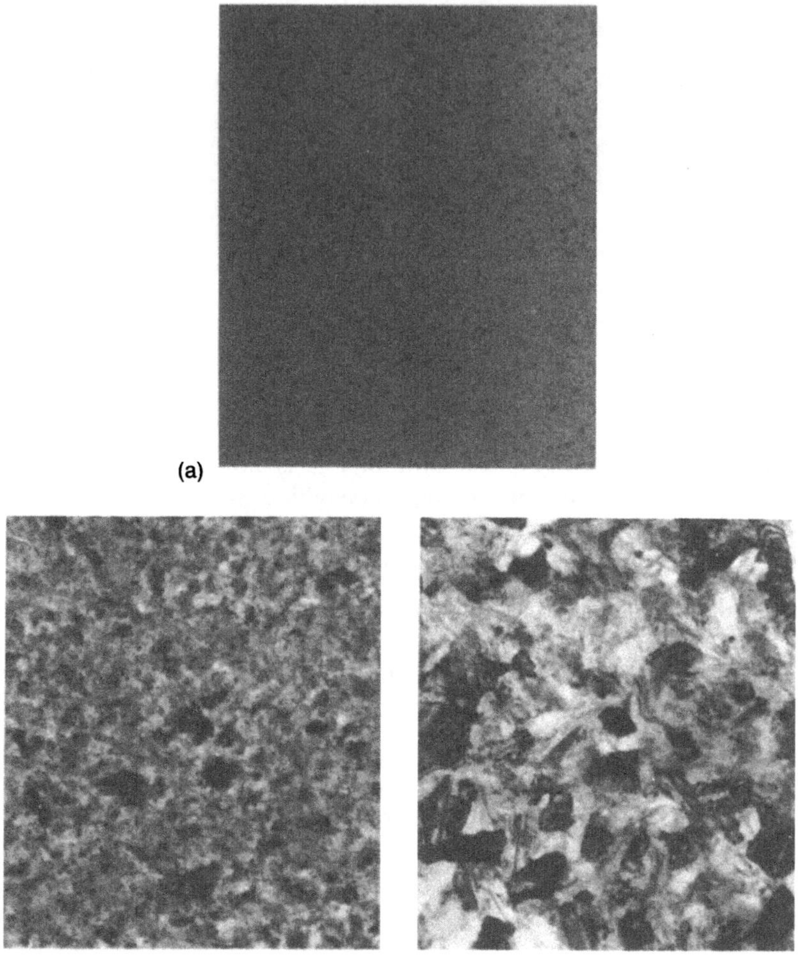

Figure 5.11 SEM image of polysilicon showing grain structures: (a) as deposited at 560°C; (b) N_2 anneal, 600°C; (c) N_2 anneal, 900°C.

rate can be enhanced by heating the substrate. From a chemical engineering point of view the situation is by no means as simple as this might suggest. If the substrate is heated electrically from the rear, the gas phase would still receive heat indirectly from the solid, causing the molecules to react in the gas phase. This is obviously highly undesirable since it gives rise to fine particles in the gas. In fact, one of the most important objectives in CVD is the control of particle formation. An example, is shown in Fig. 5.12, for $SiCl_4$. It can be seen that various chemical

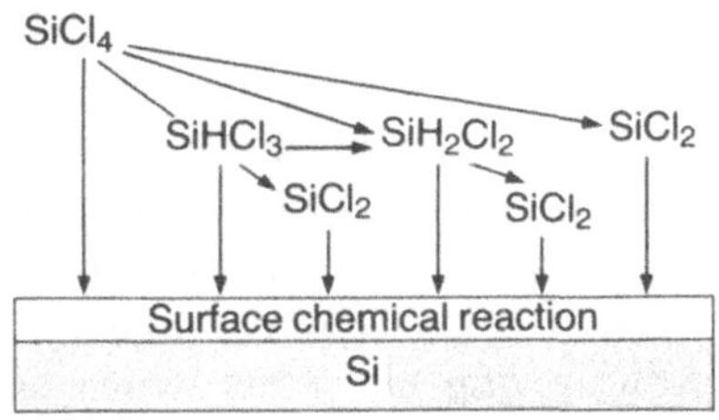

Figure 5.12 Possible chemical reactions and mass transport in CVD using $SiCl_4$ as reactant.

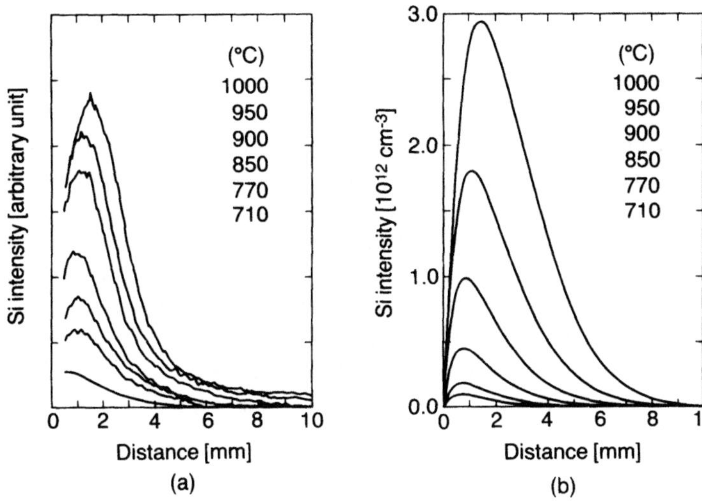

Figure 5.13 Concentration profiles of Si as a function of the distance from substrate surface: (a) measured; and (b) predicted by a model.

reactions can take place, although the relative importance of each of the elementary steps is not known.

Figure 5.13, gives an example of the kind of mechanisms involved in CVD. The predicted concentration profile of (atomic) Si using a model which takes account of chemical reaction and diffusion is compared with measured values [50]. The profile exhibits a maximum some way from the growing surface of the substrate. This suggests that atomic silicon is produced in the gas phase and diffuses to the substrate where it is adsorbed. At the same time, diffusion taking place back into the gas phase. CVD is a complicated process involving gas phase and surface reactions taking place simultaneously along with mass and heat transfer.

Plasma CVD [51, 52]
A low temperature plasma is used both in plasma CVD and in sputtering, Essentially the plasma is a gas in which electrons are accelerated by an electric field to a level such that the kinetic energy is sufficient to split the molecule into an ion and an electron which then ionizes other molecules which in turn cause further splits. The cumulative effect is an avalanche giving rise to light fast moving electrons which are accelerated by the electric field. The heavy ions and molecules move much more slowly. If the kinetic energies are interpreted in terms of temperature the ions are regarded as being cold, and the electrons as hot. Such a plasma is termed a non-equilibrium (glow discharge) plasma. The high energy electrons are important in activating reactant molecules to radicals, so that a film substrate can grow at lower temperatures than in CVD. The properties of plasma-films deposited at low temperature are generally not the same as those formed thermally. For example, the amorphous hydrogenated silicon nitride film formed by plasma CVD at 350°C is softer than the near-stoichiometric layer deposited at 750°C by CVD. These differences in the film quality are the result of the influence of temperature on the bond formation and relaxation processes, the different precursors and the effect of ion-bombardment occurring during plasma CVD (see Figs. 5.14–16). The effects of ion-bombardment are significant in plasma CVD and sputtering causes damage to the surface of the film with consequent degradation in quality. Bombarding a growing surface with ions of known kinetic energy is one way to improve film quality.

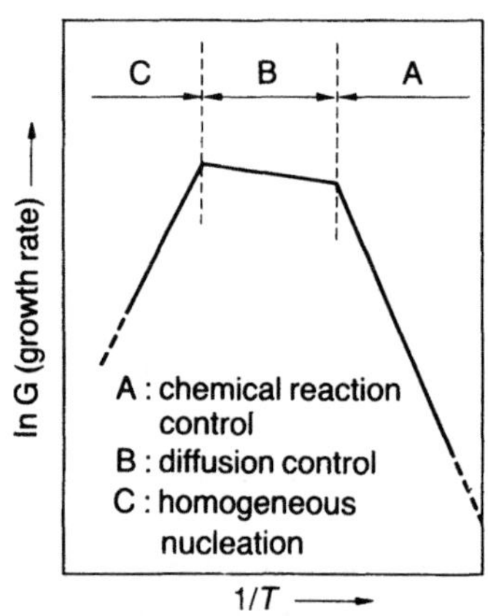

Figure 5.14 Temperature dependence of film growth rate in CVD: (A) chemical reaction control; (B) diffusion control; (C) homogeneous nucleation.

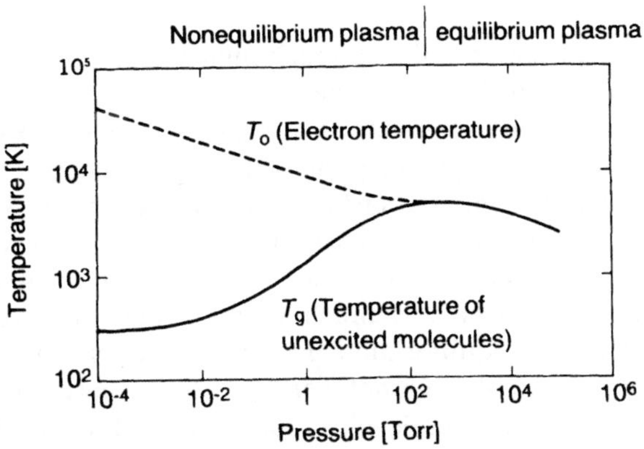

Figure 5.15 Relation between pressure and temperature of species in plasmas.

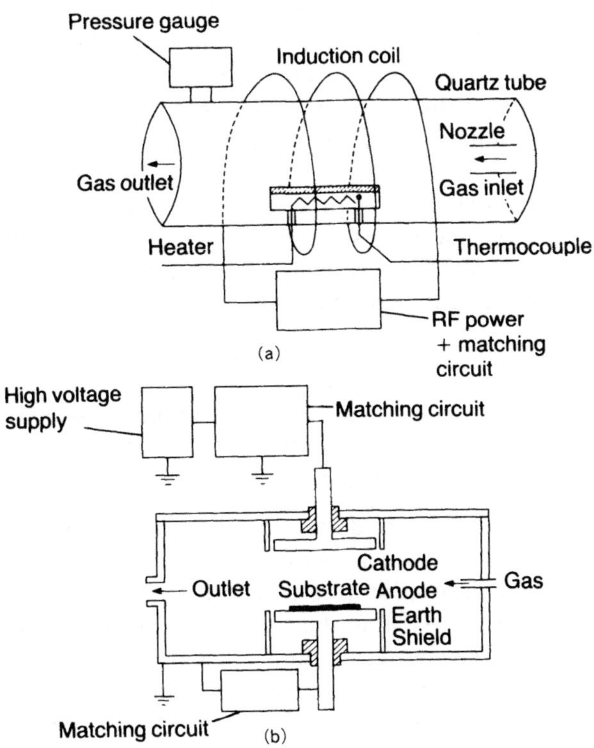

Figure 5.16 Plasma CVD apparatus: (a) inductive coupling; and (b) capacity coupling.

Sputtering [53]
A glow discharge plasma requires a much higher electric field than plasma CVD. A specially shaped solid of the chemical element used for the film is mounted in the sputter chamber. The process is analogous to throwing a ball (i.e. the ion) at a gravel path (the target surface) which causes stones to be ejected from the surface, i.e. sputtered. In VLSI, Ar^+ is commonly used as the ion and its kinetic energy is raised to 1000 eV by acceleration in an electric field. The sputtered atoms have a kinetic energy around 10 eV, which is much greater than the average kinetic energy of a thermally equilibriated gas mixture at normal temperature ($kT = 0.025$ eV). Some of the kinetic energy of the sputtered atoms is lost by the collision with gas molecules during transit from the target to substrate so that arrival at the substrate the energy is still much greater than for thermal equilibrium processes such as CVD. Kinetic energy plays an important part in determining properties such as adhesion (better) and crystallinity (worse).

Multitarget sputtering is used to produce alloy films. In addition, oxides and nitrides of metals can be deposited if O_2 and N_2 are present in the plasma. This is called reactive sputtering. The composition of the alloys and compounds is difficult to control because it relies on blending the material from the target. The difference in the sputtering efficiency and sticking probability of the elements is the biggest problem in multi-target processes. The number of atoms sputtered and the fraction of atoms colliding which stick depend on the element.

Plasma CVD and sputtering processes are similar since they are both low-temperature plasma processes, as is evident from Table 5.7. Sputtering is basically a physical process while CVD and plasma CVD are essentially chemical. These fundamental growth kinetics can be expressed as:

$$\begin{aligned}\text{Growth rate} = {} & (\text{Influx of Ar}^+ \text{ hitting target}) \\ & \times (\text{sputtering efficiency}) \\ & \times (\text{fraction of sputtered atoms arriving} \\ & \quad \text{at the substrate surface}) \\ & \times (\text{sticking probability}).\end{aligned}$$

The sputtering efficiency is determined essentially by the element and its sticking probability, which would usually be close to unity. Any enhancement in the deposition rate is therefore dependent on an increase of the flux of Ar^+ ions reaching the target, the increase in ion concentration and fraction of the sputtered atoms arriving at the substrate. Any gas phase molecules (Ar) will cause a decrease in the rate because of scattering. This means there must be a trade-off in terms of Ar pressure.

Magnetron sputtering provides one method of solving this problem, since the Ar^+ ions induced by the magnetic field near a target cause the plasma to concentrate there to give a high concentration of Ar^+ at low

Table 5.7 Comparison of plasma CVD and sputtering

	Plasma CVD	*Sputtering*
Pressure	10^{-2} ~ several Torr	10^{-3} ~ 10^{-2} Torr
Power supply frequency	1 ~ 50 W 13.56 MHz	100 ~ 300 W 13.56 MHz
Electron temperature	several eV	several eV
Growth rate	1 ~ 10Å/s	1 ~ 10Å/s
Feed method	gases (hydride, halide *et al.*)	Atoms sputtered from target by Ar^+
Species and energy	Active species like radical (SiH_2, SiH_3) $\frac{1}{40} \sim \frac{1}{20}$ eV	Atoms, clusters 10 ~ 30 eV
Major applications	Amorphous Si and alloys (a-SilH, a-SiC:H, a-SiGeH), BN, etc.	ZnO, ITO, AlN, $LiNbO_3$, TiO_2, etc.
Advantages	Multilayers by gas exchange mass production	Good operatability, wide applications including metals, semiconductors and insulators

total pressure and the scattering probability of the sputtered atoms therefore is low. This permits growth rates of 100 nm/min.

CVD, plasma CVD and sputtering are the three most important thin film deposition processes. In the case study, CVD has been used to form the SiN film by the reaction of SiH_2Cl_2 and NH_3, poly-silicon by pyrolysis of SiH_2Cl_2 and PSG by the reaction of SiH_2Cl_2, PH_3 and O_2. Plasma-CVD is used to deposit SiN_xH_y from $SiHCl_2$ and NH_3. Sputtering is used to form Al films. These are compared in Table 5.7.

A variety of other film formation processes are under development. Photo-CVD is a low temperature process using photochemical excitation of molecules. Compared with other low temperature processes such as plasma-CVD, there is less film damage because there is no ion bombardment. The growth rate is small and the high cost of the light source has restricted its use up to now. Nevertheless, it is capable of producing a high density film having a better step coverage so it still remains an attractive technique.

Improvements in plasma-CVD are still being pursued. The triode-plasma CVD causes less damage from ion bombardment by using a third electrode inserted between the substrate and the target and maintained at the same potential as the substrate. This confines the plasma between the target and the new electrode so that the ions near the substrate have

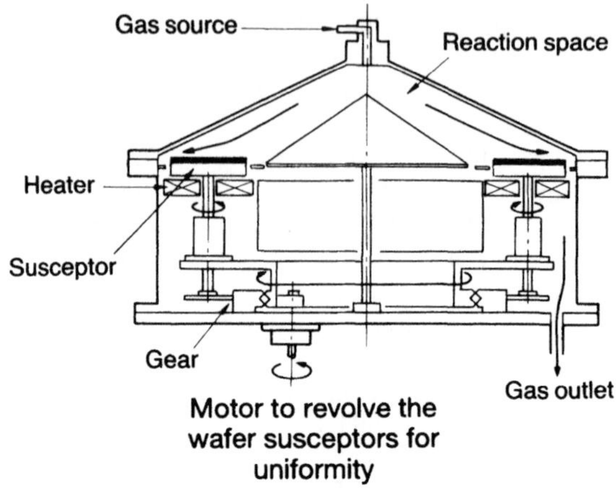

Figure 5.17 Atmospheric pressure CVD reactor.

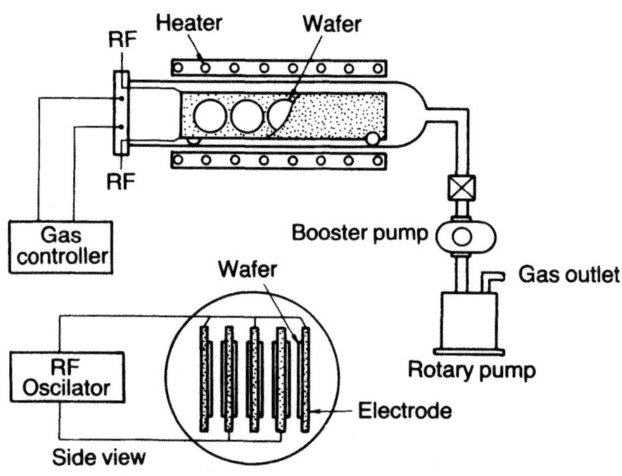

Figure 5.18 Plasma CVD apparatus.

lower kinetic energy. This approach is also used in the ECR plasma CVD method, where control is exercised by isolating the substrate from the plasma.

It is clear that there are several promising alternative processes for film deposition. The choice in a particular case has to be made on experience. A more systematic procedure will depend on a better understanding of the deposition mechanisms.

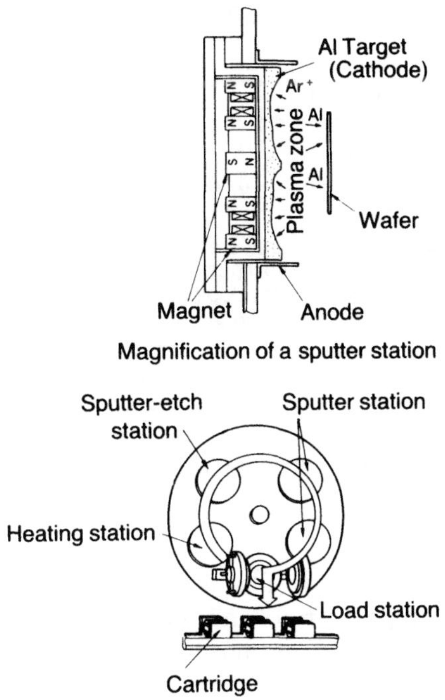

Figure 5.19 Magnetron sputtering apparatus.

Equipment and operational procedure

The major factors which need to be considered when selecting the type of equipment for film deposition has to take account of the quality of the film, growth rate, the uniformity of the film and growth rate of individual wafers and groups of wafers, control of particles, including their formation, growth and deposition as well as operability and maintenance. Different types of reactor have been developed for particular applications rather than a general purpose piece of equipment. Some of the examples are shown in Figs. 5.17, 18 and 19.

Figure 5.17 shows an atmospheric pressure CVD reactor used for depositing SiO_2 films. A lot (containing 25 wafers) is processed batchwise with one wafer being mounted on a circular substrate holder electrically heated from the rear. The holder rotates about the axis of the reactor and also about its own axis to try to keep the deposition rate profile uniform both for individual wafers as well as the batch. This is the most favoured procedure and called the rotation–revolution method. The heating system is different, as can be seen in Fig. 5.20, with the heat being supplied from an electric furnace surrounding the reactor tube. This gives

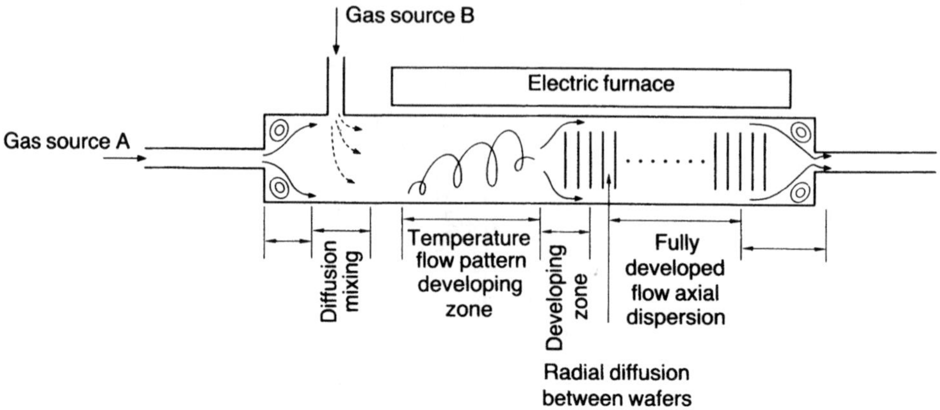

Figure 5.20 Transport phenomena in a horizontal CVD reactor.

rise to a different mode of film deposition and particle generation depending on the temperature of the reactor wall is. The reactors in Fig. 5.17 and Fig. 5.20 have cold and hot walls respectively.

Figure 5.18 shows a plasma-CVD reactor for a batch of 25 vertically mounted wafers placed between plate electrodes. The reactor is classified as being capacitance-coupled with a facing target (see Fig. 5.16 [54]). A problem with this arrangement of electrodes is the tedious job of loading and unloading the wafers.

Figure 5.19 shows a typical magnetron sputtering reactor. The growth rates using this type of reactor range from 10 to 100 μm/min, over a wide range of temperatures and pressures. The process time required to obtain a film of 1 μm is between 10 and 100 minutes. Productivity depends on the time, it tends to be compatible with the times for preheating, loading, cooling down and unloading. If a good quality film can be formed in 10–100 min, then this would certainly be acceptable.

With increasing size of wafers, it is possible to consider individual processing rather than in lots of 25. A move from batch to semi-continuous (so-called leaf-to-leaf) processing places greater emphasis on higher growth rates.

Transport phenomena in the reactor

Simultaneous reaction and transport [54]

Deposition can be carried out in a variety of ways e.g. CVD, plasma-CVD and sputtering, each one having a distinctive set of operating conditions. The way in which gas flow, diffusion, energy transport, chemical reaction, plasma chemical reactions and sputterings interact to determine the rate-limiting steps and the quality of the film depends on the particular technique being used.

An example of a horizontal-type of CVD reactor heated by an external furnace is shown in Fig. 5.20. Two gas streams of SiH_2Cl_2 and NH_3 are premixed in the cold zone of the reactor, which allows the temperature and concentration profiles to develop. In this region natural convection tends to predominate. The batch of wafers is located where the profiles are fully developed. In some cases it may be necessary to place several at the front of the batch to act as a guard to ensure that the conditions experienced by the batch are uniform. Because the molecules diffuse from the main stream into the space between wafers, and deposit on the surface, inclining the wafers at an oblique angle to the reactant gas flow can affect deposition profiles since the chemical reaction takes place in the gas phase as well as on the surface. Controlling the operating conditions to ensure uniform deposition is therefore important and there is a need to have a clearer understanding of the mechanism of the chemical reactions.

Generally, it is the transport phenomena which govern the uniformity of thickness and quality of the film. The concentrations of the reactants decrease in the direction of gas flow so that a constant temperature would cause the film thickness to decrease. To compensate for this requires an increase in temperature. For a tolerance of $\pm 1\%$ in the film thickness in the case of polysilicon from SiH_4 (a first-order reaction with an activation energy of 200 kJ/mol) the concentration needs to be controlled to within $\pm 1\%$ and temperature to a fraction of a degree in order to ensure the growth rate is maintained to within $\pm 1\%$. This is clearly a demanding control problem.

Using temperature profile to compensate for changes in concentration so as to maintain the growth rate will affect the quality of the film. In the case of silicon nitrides for example, temperature influences both the stoichiometry and density of the film. The temperature profile must therefore be kept within strict limits to avoid this. Such information is not generally available, because of the commercial sensitivity, but the effective usage of molecules is likely to be less than 20%.

The reactant concentration varies over the radius of the wafer and this is responsible for non-uniformity in the film since the balance between deposition rate and diffusion changes. A low rate of diffusion results in a concave film with the thickness near the centre being thinner than that at the edge. The diffusion rate can be increased by increasing the spacing between wafers, but then the productivity per unit volume of reactor is reduced, so there is scope for optimization.

The mathematical model to describe this involves the Navier–Stokes equations in conjunction with the continuity equation and material and energy balances. To simulate the whole process requires quantitative

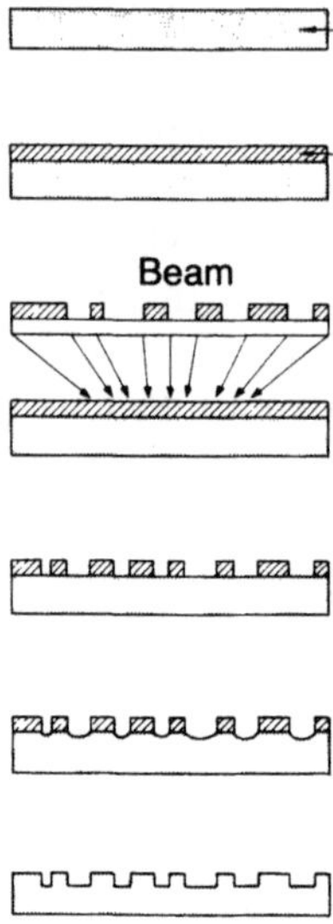

Figure 5.21 Lithography and etching processes.

knowledge of the chemical reactions and deposition process. Recent work has been directed to understanding such simultations, in order to find optimal processing conditions.

Transport processes at reduced pressure [55]
Many film preparation processes are operated at reduced pressures, including all the glow discharge plasma and some of the CVD processes. This is quite different from the petrochemical industry and requires special analysis of which operations at reduced pressure make it easier to achieve uniform films. The diffusion coefficient is inversely proportional to pressure while viscosity and thermal conductivity are unaffected. This could be anticipated by noting that mean free path increases from the order 0.1 micron at an atmospheric pressure but is about 10 cm at 10^{-3} torr, which is of the same order as the size of the reactor. Under these conditions, the flow characteristics progressively change from viscous to Kundsen (molecular) flow.

5.2.5 Lithography

Basic principles

The process of mapping parts of a circuit pattern in the form of a mask on the surface of a silicon wafer is called lithography. It is similar to the printing of a photograph. In the first step, the whole surface of a wafer is coated uniformly with a thin layer of resin called a photo-resist. This is then exposed to light which is projected through a mask having the pattern which it is desired to print. A photo-reaction takes place in the resist over the exposed area. The part of the resist where the photo-reaction has taken place (or where the resist was not exposed to the light, depending on the type of the resist) is removed using developing agents and the pattern is transferred to the wafer surface (Fig. 5.21).

This process is followed by etching or ion implantation and may be repeated several times in the VLSI process. It is used to form a desired pattern on the wafer after etching, as described in the next section. In some cases, it is used to cover parts of the wafer surface with a resist to allow ion implantation in a region and form a trench.

To describe the ultra-high levels of integration required in fabricating VLSI devices using super-micromachining techniques, it is usual to characterize the scale (size) by the design rule. For the fabrication of a 1Mbit DRAM it is 1.3 microns, decreasing to 0.8 microns for a 4Mbit DRAM. VLSI calls for strict process management and control particularly with respect to lithography and etching in order to work to these limits.

Equipment and materials

In addition to photomasks and photoresists, lithography requires exposure machines which can function as steppers as well. Usually, the photomasks are quartz plates having circuit connections (patterns) of chromium. The dimensions of the pattern are either actual size or 5 to 10 times larger. For a 1Mbit DRAM chip this is typically 7 mm × 14 mm. Masks having magnified patterns are called reticles.

The quartz plates are about 10 cm square and are flat to within 2 microns. The masks are designed using CAD techniques and fabricated by electron beam (EB) drawing. The lithography and dry etching processes create a pattern on a quartz plate which has had an 80 nm thick layer of chromium deposited on it. Electron beam machines are now capable of drawing patterns having a minimum line width of 0.1 microns to an alignment accuracy of 0.04 microns.

The number of masks required will correspond to the number of lithographic stages. In the case study, for example, thirteen lithographic processes are involved so at least thirteen different masks need to be prepared. Nowadays, the masks are not put in contact with the wafers during exposure because projection printing is used, so they can be used indefinitely. However, the continual improvements in circuits means that the masks are always changing.

Dust particles gives rise to catastrophic defects during spreading of the photoresist and exposure. Consequently, the air used in such process

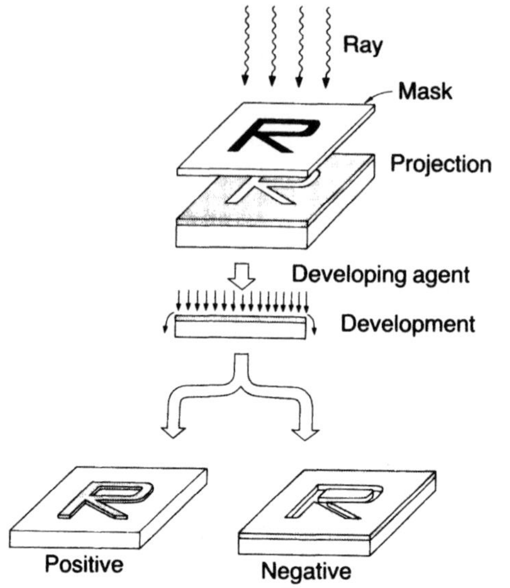

Figure 5.22 Pattern formation.

environments should be ultra clean (class 10). In addition precise temperature control is necessary to ensure accurate alignment.

Photo-resists

Detailed descriptions of the photo-resists used in lithography are described in detail elsewhere (section 6.2.1), so only a brief outline is given here. They consist of a base polymer and photosensitive agent. On exposure to the UV rays the photoactive agent reacts to give either a photo-decomposition product which is soluble in the developer (positive resist) or to a cross-linked polymer which is insoluble (negative resist). The patterns formed on the wafer surface by the two types of resist represent opposite images, as illustrated schematically in Fig. 5.22.

Up to 64Kbit DRAM, negative resists have been used. However, there has been a move to positive resists for VLSI because they are capable of higher resolution. Also, since the sensitivity of the resist is a function of the wave length of the light there is a need to match these as well.

Resists such as Navalc resin (napthaquinonazido) are used for the production of 265Kbit DRAMs because they are capable of yielding the higher sensitivity required at this level of resolution. The resist reacts on irradiation by UV light to give a material which is soluble in alkaline solution according to the reaction sequence:

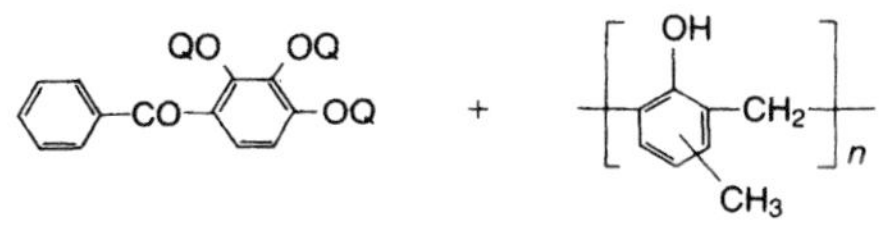

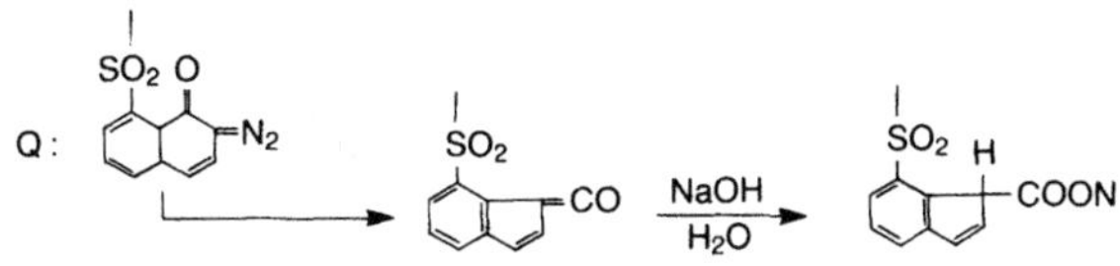

It is important that the resist should have strong resistance to the etchants used after lithography and have very good contact with the wafer surface.

Coating

The wafers are held on a vacuum chuck and the resists are mixed with solvents to adjust the fluidity and then sprayed on to the surface. The chuck is then rotated at a speed in the range 2000–5000 rev/min to remove excess material and to form a uniform coat about 1 micron thick. This is called the spin-coat method. It is possible to control the thickness to between 0.5 and 5 microns by adjusting the rotational speed and/or the viscosity of the resist solution. The resist layers are then dried (soft baked) by heating in an oven at 80°C for 15 minutes. The goal is to produce a uniform coat, free of defects such as pinholes. This is not easy especially when the surface is uneven because of the patterns formed in earlier stages. To overcome this, multi-layer resist techniques can be used, as will be discussed later.

Exposure

This stage in lithography involves the transfer of the mask pattern on to the photo-resist by UV or deep UV irradiation. This may be done directly or at reduced magnification. The light source for the UV is usually an ultra-high pressure mercury or xenon lamp: deep UV exposure xenon-mercury lamps are used. Up to 4Mbit DRAM, UV rays having a wave length of 436 nm are used. They give a resolution of 0.6–0.7 nm. For higher resolutions, shorter wave lengths, such as X-rays or electron beams have to be used.

Aligning the silicon wafers with the mask is done automatically or optically by lining up reference marks. Improved resolution can be achieved using reduction projection which involves combination of a reticle and a stepper. The reticle has about four circuit patterns printed at a magnification of 5 or 10 times larger than the actual size. The patterns are then printed on the wafer by projecting an image on the surface and

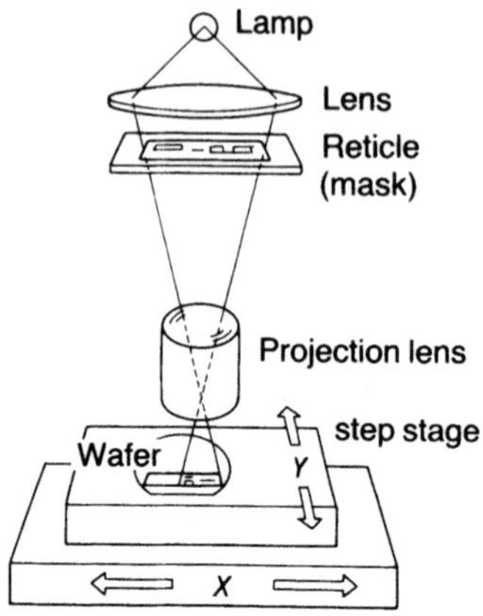

Figure 5.23 Size reduction projection printer.

the whole process is repeated. The wafers move in a two-dimensional plane so that various parts of the wafer can be exposed. Average exposure time is a few seconds so the throughput is relatively slow, typically 30–60 wafers per hour. The accuracy of the alignment is to within ±0.3 microns. A schematic diagram of a step and repeat exposure machine is shown in Fig. 5.23.

Developing

After exposure, the developing is carried out using special solutions. In general alkaline solutions are used for positive photo-resins, and organic solvents such as xylenes, for negative resins. The solution is sprayed over the wafer and dissolves part (or not, depending on the type) of the resin which has been irradiated by the UV rays. The resolution and sensitivity of the pattern depends totally on the developing process, so that close control of the concentration and temperature of the liquid, in addition to developing time, is essenital. With a positive resist, developing generally requires 60–90 seconds, after which it is washed with water.

Inspection is carried out under a microscope to check the completeness and the uniformity of the developing. This can be done either by eye or automatically. If developing is incomplete, the wafers need to be returned for further treatment or even to the resist spraying process. In the latter case the previous layer must be removed first.

Hard-baking

The last stage involves (hard) baking at 150–200°C for about 30 minutes to remove the last traces of solvent in the resin. It also serves to thermally

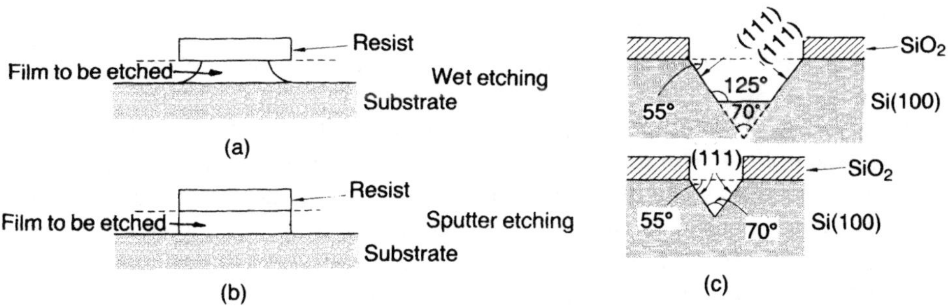

Figure 5.24 Schematic representation of isotropic and anisotropic etching processes: (a) isotropic; (b) anisotropic; and (c) anisotropic etching for (100) silicon waters in alkaline solutions.

treat the unhardened part of the resist and improve the resin-wafer contact as well as the resistance of the resin to etchants in further stages.

5.2.6 Etching

Lithography leaves a resist pattern on the silicon wafer which is exactly the same as that of the mask. The exposed areas are then etched with suitable chemicals, after which the resist is removed using appropriate solutions. Thus, etching can be regarded as a method of transferring the pattern by chemically dissolving or physically evaporating part of the silicon or silicon oxide layer. It can be done under either wet or dry conditions using liquids or gases respectively. Dry etching has the advantages of being easier to control with no requirement to treat waste liquid and it is not necessary to monitor end points, so is well suited to automation.

Another method of classifying etching is as isotropic or anisotropic. Dry etching with plasma and wet etching in an acid solution are isotropic processes. Wet etching in alkaline solution is anisotropic. Reactive ion etching (RIE) is anisotropic, even though it is carried out in the gas phase, as will be seen later. Isotropic etching proceeds uniformly in all directions so that part of the substrate under the mask will also be etched giving rise to hillocks, as shown in Fig. 5.24(a). On the other hand, anisotropic etching proceeds in two different ways: by sputtering, as in RIE, which gives sharp patterns because the etching is normal to the surface; or with alkaline solutions, where the etch rate depends on the orientation of the crystallographic planes, as illustrated in Fig. 5.24(c). This shows how (100) wafers have been etched in aqueous alkaline solutions to produce trenches bounded by (111) faces.

Wet etching

Aqueous HF solutions containing oxide agents are suitable for isotropic etching of Si and are used to ensure a flat surface. At the present time, alkaline solutions composed of either EPW (ethylenediamine + pirocatechole + water) or KOH are used to form isolation structures because they have better anisotropic features. In the latter cases, SiO_2 layers can be used as masks and etching results in the generation of hydrogen bubbles. The etching rate is a function of the concentration and temperature of the liquid but can be as high as 100 microns/hour. Agitation is necessary to ensure a uniform rate over the surface.

Dry etching

Because it has more advantages, it is now usual to use dry etching. In the case study, SiO_2, Si_3N_4, polysilicon, PSG and Al will be considered. The etchant gas reacts with these materials to form volatile substances which are removed, so allowing the etching to proceed. Removal of the resist after lithography can also be carried out by dry etching. In practice, plasma and sputter etching are used, the latter being anisotropic. Sputter etching can be further divided into chemical and physical as can be seen from the list in Table 5.8. Some of the points are discussed in more detail below.

Plasma etching

Plasma is generated when a R-F source is applied to a stream of CF_4 containing a few per cent of oxygen in the pressure range of 0.1 to 1 torr. This produces fluoride radicals $F^{\cdot}$, which in turn react with the silicon wafer to form (volatile) SiF_4 which is then purged. The reaction can be written as follows:

$$CF_4 = C + 4F$$
$$Si + 4F^{\cdot} = SiF_4$$
$$SiO_2 + 4F^{\cdot} = SiF_4 + O_2$$

Isotropic plasma etching can be achieved by arranging the silicon wafers so they are perpendicular to the axis of the reaction chamber. The etching rate for a particular material is independent of the gas composition, temperatures and gas flow rate and is in the order:

$$\text{polysilicon} > Si_3N_4 > SiO_2$$

However, the actual rates do depend on the conditions, the location of wafer and the size of the equipment. The rate for Si_3N_4 is typically in the range 10–80 nm/min.

In parallel plane plasma etching wafers are located on the anode side in

Table 5.8 Comparison of dry etching methods [107]

Type	*Plasma etching*	*Parallel plate plasma etching*	*Reactive sputter etching*
Outline	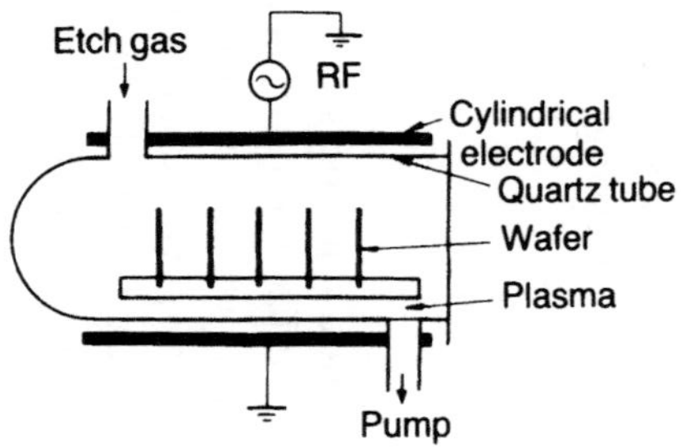	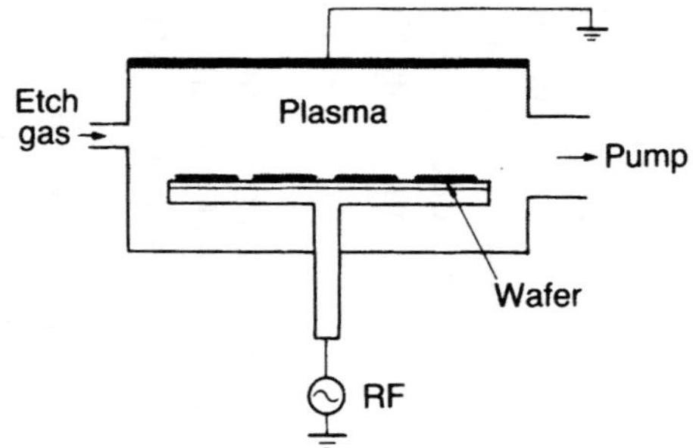	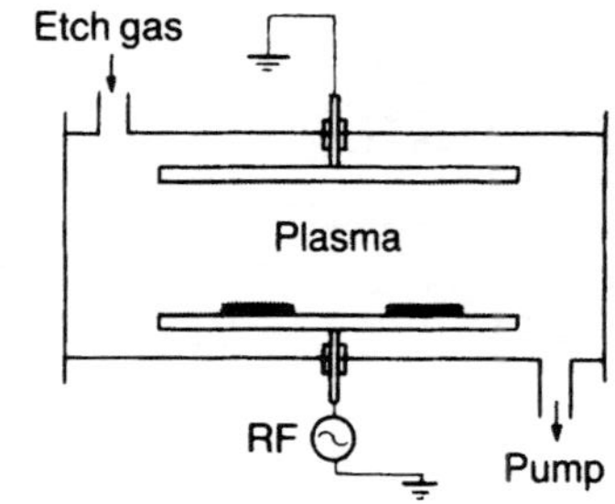
Wafer location	Outside of plasma area	Inside of plasma area	Inside of plasma area
Gas pressure	$10^{-1} \sim 1$ torr	$10^{-1} \sim 1$ torr	$10^{-2} \sim 10^{-1}$ torr
Reaction species	Radicals	Radicals and few ions (with slight sputtering effects)	Radicals and ions (chemical reaction and sputtering effects)
Etching pattern	Isotropic	Isotropic/anisotropic	Anisotropic
Materials to be etched	Polysilicon/silicon Si_3N_4/photoresist PSG	Polysilicon/Al SiO_2 on Si PSG	Polysilicon/Al SiO_2 on Si
Problems	Undercut (side etching) Not applicable to SiO_2 on Si Applicable to Al	Undercut (side etching) Poor resistivity of photoresist Radiation damage	Corrosion of photoresist Radiation damage

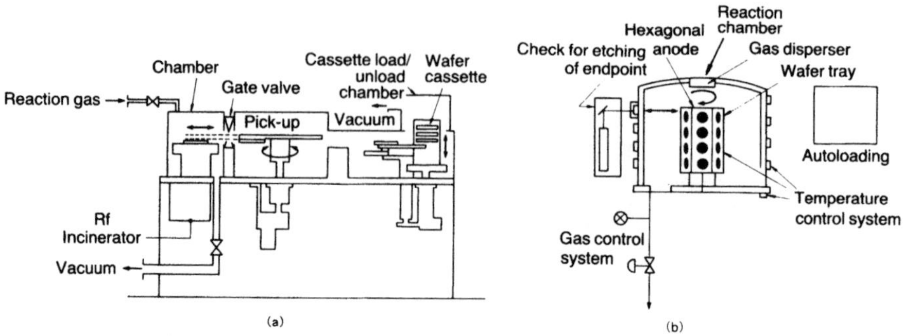

Figure 5.25 Etching apparatus: (a) single water etching; and (b) batch-wise etching.

parallel plane plasma etching, so fluoride radicals and some of the ions attack the surface normally causing sputtering effects and anisotropic etching. This arrangement has the advantage that the temperature rise in the wafers can be kept down by cooling the electrode. Both batch and single wafer processing equipment are available commercially (Fig. 5.25).

Reactive ion etching (RIE)

Because the etching rates of silicon single crystals and silicon dioxide are equal in plasma processes, RIE can be used to etch the silicon dioxide layer of silicon wafers. By applying an R-F signal at 13.56 MHz and at pressures between 0.01 to 0.1 torr, ions other than radicals are produced and they impinge perpendicular to the wafer at the cathode which results in anisotropic etching. The rate depends on the gases and equipment. A typical growth rate of thermally generated oxide layers is reported to be 600 Å/min. Both batch and single wafer processing equipment are available.

Detection of end point

At the end of the etching process, the reaction must be stopped and the resist removed. A plasma asher is used to oxidize the resist in an oxygen atmosphere and decompose it. Alternatively, when wet etching is used the resist is removed by dipping the wafer in a special liquid which causes the resist to swell so that it can then be easily removed. Positive resists which have been baked at temperatures less than required for hardening are removed by alkali solutions after irradiation with the UV light.

6

Packaging and process monitoring technology

6.1 PACKAGING

After the wafer processing finishes, the next stage is packaging. This involves cutting a fabricated Si wafer into chips and connecting the bonding pads of the chips and metal lead frames with fine wires. To prevent the chips from severe environments, they are covered with passivated materials. At this stage, it is possible to make the electrical connections to the printed circuit boards. The packaging technology necessary to seal Si devices is explained below.

6.1.1 Background

An outline of the history of the packaging is shown in Table 6.1 [56]. Metal can seals were originally used for Ge transistors but these were replaced by adapting the method used for Si transistors, which became possible as a result of improved passivation of Si chips. Plastic packages for IC, LSI and VLSI have replaced hermetically sealed systems because of reduced cost and ease and mass production. Figure 6.1 shows how high density packaging technologies have developed very rapidly in respect of making:

- the packages smaller (from DIP to SOJ);
- thinner (e.g. ZIP, QFP and TSOP);
- surface mount devices (QFP, SOJ, TSOP, TAB etc.); and
- a variety of packages (QFP, PGA and others).

For very common devices such as memory, the package shape needs to be standardized without increasing the number of pins significantly. This has been achieved despite the density of the Si chip increasing by a factor of four over three years. At the same time, packaging technologies have

Table 6.1 History of semiconductor devices and their packages

Age			*1950*	*1960*	*1970*	*1980*
Device		Ge-Diode	Si-Diode Ge-Tr	Si-Tr	IC LSI	VLSI
Package	Diode / Tr		Can	Plastic (potting)	Plastic (transfer)	
	IC / LSI / VLSI			Can	Plastic DIP (transfer)	Plastic SMD (transfer)

SMD: Surface mounts device [56]

been adapted for such high density mounting. On the other hand, logic devices such as gate arrays and microcomputers have tended to result in an increase in the number of pins so as to achieve greater functionality, as shown in Fig. 6.2 [57].

The main problems associated with plastic packaging are indicated in Table 6.2 [58]. For high density mounting, the following problems must be overcome:

- reduction of damage to the high performance Si devices due to greater integration and higher density; and
- balancing of the stresses between the semi-conductor material, lead frame metal, organic resin and chip bonding material.

This is because the volume ratio of Si chip to package becomes larger and approaches 1. Consequently, soldering of the joints on the printed circuit becomes critical. The approach to estimating the thermal stresses which arise from bonding the various materials is indicated in Fig. 6.3. To reduce the thermal stress, it is necessary to match the thermal expansion coefficients of the materials and ensure that these are compatible with the elasticity of the encapsulating resin.

There are also technical problems in package structure design arising from miniaturization, as shown in Fig. 6.4. Mounting problems become more critical as Si chips become larger and require special techniques to overcome them. The reliability of plastic sealing is approaching that of hermetic encapsulation. This advance has been brought about by improving the multi-level interconnection structure and passivation techniques in the water fabrication process. The development of highly

Type	Package	Outline	Package material, pin count (pitch)
Pin insertion type	DIP (dual-in-line package)	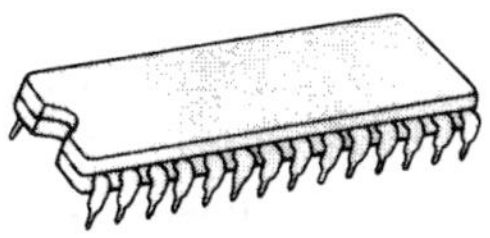	Plastic, ceramic 18–64 pin (2.54 mm)
	ZIP (zigzag in-line)	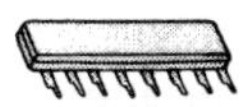	Plastic 16–40 pin (2.54 mm dual row)
	PGA (pin grid array)	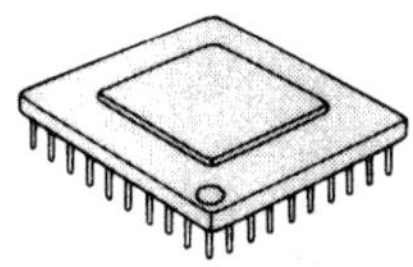	Ceramic 40–600 pin (2.54 mm)
Surface mouthing type	QFP (quad flat package)		Plastic 20–300 pin 1.0, 0.8, 0.65, 0.5 mm) (quad row)
	SOJ (small outline J-lead package)	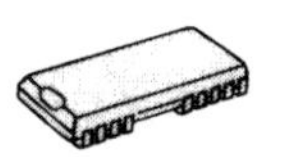	Plastic 20–28 pin (1.27 mm) (2 Dual row) (J leaded)
	TSOP (thin small outline package) TAB (tape automated bonding)	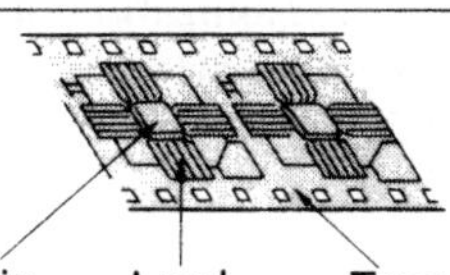 Chip Lead Tape	Plastic 20–32 pin (0.5 mm) Plastic 8–1000 pin (0.1–1 mm)

Figure 6.1 Packages for VLSI [56].

reliable plastic sealing has resulted in large cost savings and played an important role in expanding the use of VLSI.

When surface mount packages are placed on a printed circuit board, they are exposed to temperatures up to 200–250°C. At these conditions the bonding strength of the resin is reduced by about an order of magnitude. Resin materials absorb water vapour which then evaporates rapidly giving rise to high stresses. In some cases, the stress can crack the packages. It is important that the thermal stresses in the following four materials are well-balanced.

1. Si components of the semiconductor: thermal expansion coefficient $\alpha = 3 \times 10^{-6}$/°C.
2. Metal for lead frame: new materials e.g. Cu alloy $\alpha = 17 \times 10^{-6}$/°C, Fe–Ni alloy $\alpha = 10^{-5}$/°C. These expansion coefficients are close to

Table 6.2 Main technology problems of plastic package structures [58]

Problem		*Countermeasure*
Reduction of device damage	Optimum design of package structure	Stress analysis simulation
	High adhesion, low stress system	Low stress, high adhesion resin system
	Reduction of impurity and ion concentration	Reduction of Cl ion, U concentration and other impurity
Leadframe	Metal of high adhesion with mold resin	Improvement of resin adhesion
	Minimization of partially plated area	Improvement of accuracy of lead-frame plating
	Stress reduction of lead cutting	Stress behaviour analysis of cutting
Solder joint	Stabilization of solder joint	Improvement of lead fabrication accuracy Stabilization of lead plating film
	Long-term stabilization of solder joint part	Behaviour analysis of solder joint alloy
	Diversification of soldering method	Condition setting of package reliability maintenance
Environment	Anti-humidity	Package structure of high adhesion
	Thermal resistance	Design of heat balance Thermal resistant resin
	Thermal resistance to soldering	Behaviour analysis of thermal resistance to soldering

that of the plastic resin. For the Fe–Ni alloy previously used the coefficient was 5×10^{-6}/°C.

3. Plastic resin: the epoxy resin has a low expansion coefficient of about 17×10^{-6} to 10^{-5}/°C. The resin previously used had a coefficient of 24×10^{-6}/°C.
4. Organic material for chip bonding: the material to bond the Si chip to the lead frame has been changed from an Au–Si alloy to a low stress Ag or rubber paste because it is cheaper.

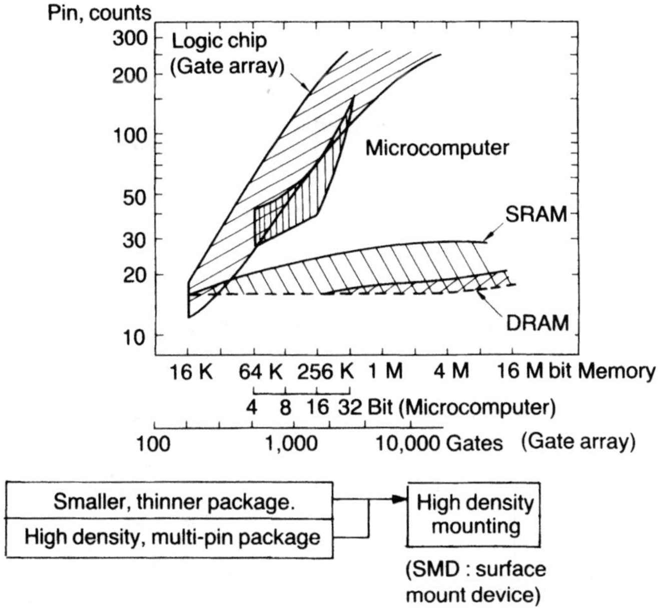

Figure 6.2 Trend for package pin counts [57].

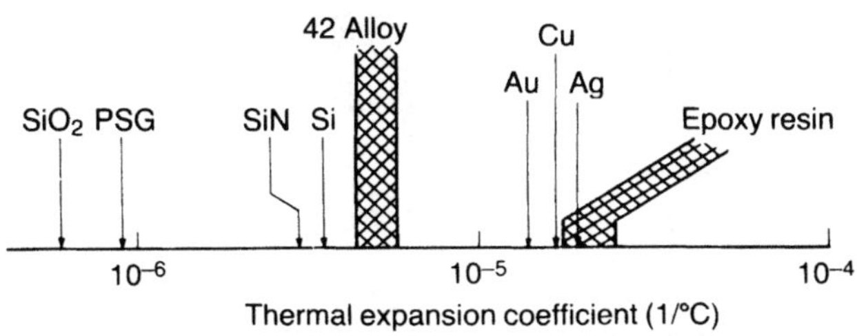

Figure 6.3 Package structure and the thermal expansion of its component materials. Thermal stress: $\sigma = \int E(T) \times \alpha(T)\, dT$; E = elastic modulus; α = thermal expansion coefficient; $\sigma = k\{(\alpha_g - \alpha_s)\, E_g(T_g - T_1) + (\alpha_l - \alpha_s)\, E_l(T_2 - T_g)\}$.

6.1.2 Assembly

The assembly process is shown in Fig. 6.5.

1. *Dicing*: After completing the wafer processing, the wafer is cut into Si chips by a high speed rotary diamond blade.
2. *Chip bonding*: Each chip is attached to a lead frame. This operation is carried out using AuSi having a eutectic point of 370°C. It is formed by

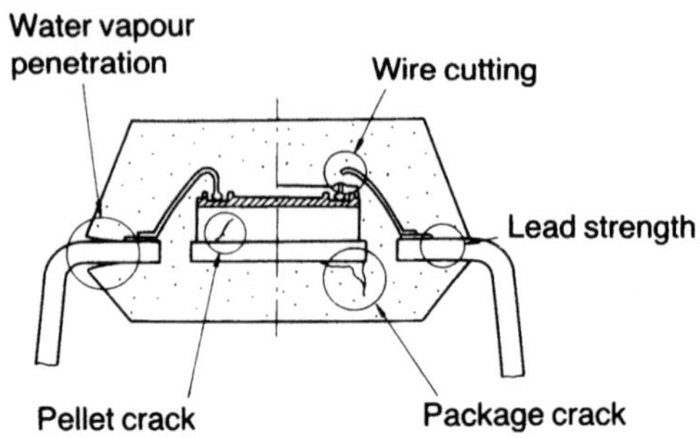

Figure 6.4 Technology problems of package structure design.

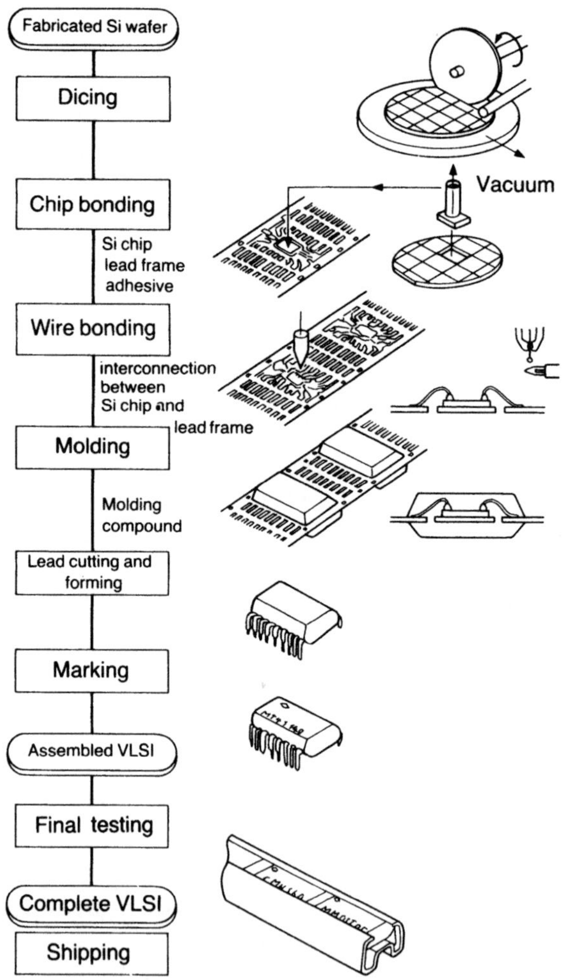

Figure 6.5 Assembly process of VLSI.

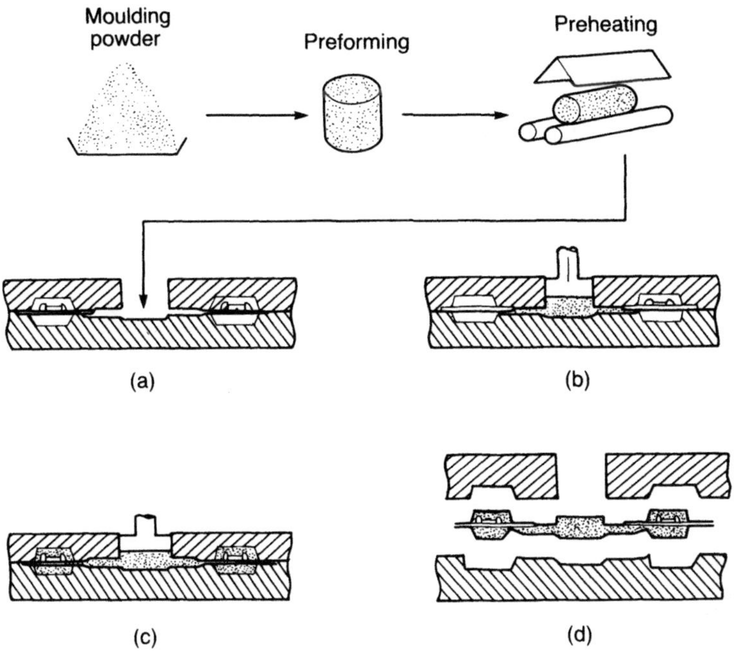

Figure 6.6 Transfer moulding process [56]: (a) clamping and placing in the mould; (b) transferring; (c) pressuring and curing; (d) mould opening and package removal.

reaction between an Au-plated lead frame and a heated Si chip. Au has been replaced by Ag or rubber paste because Au is very expensive.

3. *Wire bonding*: Wire bonding is carried out to connect the Si chip to the lead frame. In the 1960s, this operation was done using a microscope by skilled production workers. In Japan, automatic wire bonding machines have been developed using refined pattern recognition techniques which are capable of working at very high speeds. It takes approximately 0.15 s to connect a bonding pad and a lead frame. Satisfactory electrical conductivity is achieved by wire thermo-compression binding to a 30 μm diameter Au wire.
4. *Plastic moulding*: Packages are made of resin and a wire-bonded Si chip is sealed in the thermo-setting resin and the mould is hardened by heating to form the final package in a die. Figure 6.6 [56] shows the most commonly used transfer moulding process for encapsulating VLSI devices. The moulding dies have cavities for 40 to 200 packages. The encapsulation material is first preheated in the equipment by high frequency induction heaters and is then injected into the feed vessel of the die which charges the material into the cavities. After curing, the die is opened and the packages removed.

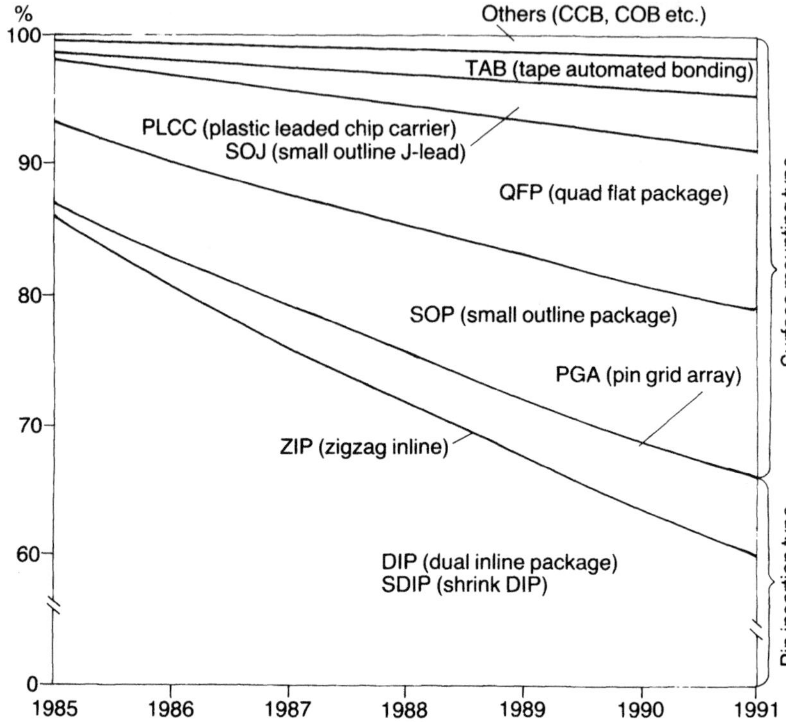

Figure 6.7 Trend for package configuration [57].

6.1.3 General overview

Figure 6.7 [57] shows how high density packaging has changed to achieve higher performance and capacity but with smaller sizes. It should be noted that the use of pin-insertion has decreased and surface mount increased. Memory devices based on larger chips need to be based on standardized packaging since high density mounting is required. In microcomputers, however, logic devices need thin packages with multiple pins.

The specification for the epoxy resin used for encapsulation has become extremely demanding, as shown in Table 6.3 [56]. In order to ensure minimum stress in the package, the following factors need to be satisfied:

- the material should have a low thermal expansion;
- the Cl concentration should be reduced to prevent Al corrosion; and
- the concentration of U and Th should be low to prevent 'soft error' problems.

The ability to be able to predict good estimates of thermal stress will make it possible to design robust packaging which will greatly increase the range of application of VLSI devices.

6.2 PROCESS MONITORING TECHNIQUES

6.2.1 Process measurement

The measurements used in semiconductor manufacturing can be grouped broadly into in-process and off-line evaluation. In Figure 6.8, the variables which need to be monitored online together with the appropriate techniques are listed, for both pilot-scale and mass-production. At the development stage of the unit processes, various techniques are used.

Online monitoring is used to check the process parameters after photo-

Process	Evaluation issues	Evaluation methods
Oxidation	Thickness Crystal defect Electrical characteristics (breakdown voltage, surface state, leakage current)	Ellipsometer · optical-interference method TEM, X-ray diffraction, SEM, Raman spectroscopy C-V method, C-t method, BT
Ion implantation Diffusion	Sheet resistance Diffusion depth Impurity concentration profile Crystal defect	Four-point probe technique SIMS, spreading resistance, SEM SIMS, spreading resistance, C-V method SEM, TEM
Dielectric film Chemical vapor deposition (CVD)	Thickness Refractive index Composition · impurity Coverage	Ellipsometer, optical-interference method Ellipsometer X-ray fluorescence spectroscopy, chemical analysis SEM
Photolithography Etching	Resist thickness Pattern size Processed form Pattern inspection	Ellipsometer, optical-interference method, mechanical stylus method Size measurement, SEM Size measurement, SEM, TEM Optical microscope, SEM
Metallization	Thickness Composition Coverage Internal stress Crystal structure	Sheet resistance, mechanical stylus method Chemical analysis SEM Hardness tester SEM, TEM, X-ray diffraction
Process integration	Particle contamination Impurity contamination Clean room Pure water Gas · chemicals · materials	Wafer particle detection SIMS, AES, Chemical analysis, chromatography Particle counter Relative resistivity Chromatography, chemical analysis

Figure 6.8 Technical issues necessary for the evaluation in the VLSI manufacturing production line and relevant evaluation techniques.

Table 6.3 Trend in epoxy encapsulation materials for transfer molding

Year		*1970*	*1975*	*1980*	*1985*	*1990*
Levels of integration (DRAM)			16 K	64 K 256 K	1 M	(4 M)
Requirements			Flame retardancy	High purity Low stress	Low α-ray	
Technologies		Anhydride cure	Phenolic cure		Silicone modification	Synthetic filler
Physical properties	Thermal stress (MPa)	12	8	4–5	3–4	<2
	Thermal expansion coefficient $\times 10^5$ (K^{-1})	2.5	2.2	1.9	1.7	<1.4
	Hydrolyzable Cl content (p.p.m.)	1000			100	<100
	U content (p.p.b.)		100		10	<1
	Flame retardancy	UL-94 HB		UL-94 V-O		

Productivities	Curing time (s)	120–150	120	90	60–90
	Minimum melt viscosity (Pa s)	10–30	30	40	40–50
Reliabilities	Moisture resistance* (Al corrosion)	1	5	10	20–30
	Temperature cycling* (package crack)	1	10	100	500

* Reliability comparison on the same TEG chip

lithography and etching, metallization and diffusion and to control the production line. This includes monitoring equipment to check that quality is maintained as well as ensuring the yield is kept at a high level. Such online information provides the necessary feedback to enable a stable and high yield to be achieved. The continuing advances in VLSI have resulted in ever-higher limits on manufacturing tolerances. To meet this, it is essential to have a reliable production strategy. High accuracy measurements are crucial if the production line is to be precisely controlled. Offline measurements can be sub-divided into the characterization and determination of the structure at the development stage and the analysis of failures on the production line.

In-process measurement requires inspection to be carried out while the wafers are moving. This means that a no-contact and non-destructive inspection technique is necessary. The usual procedure is to examine samples taken from the manufacturing line, then return them to the line. If all wafers are inspected, the measurement technique should be non-contact and non-destructive. Alternatively, special wafers can be included in the line which are used for inspection (test wafers). They are included as part of the production line in order to provide an indication of process conditions. This section is devoted to a description of the methods commonly used in semiconductor manufacturing with particular reference to in-process inspection.

As a first step, it is useful to set the problems in context. The prime objective of VLSI manufacture is to ensure that a prescribed circuit pattern is achieved, whatever the scale of working. In seeking to miniaturize the circuits, the limit of optical techniques has been reached and finer spatial resolution is only possible with electron beams from scanning electron microscopes.

With smaller and smaller scales as the level of miniaturization increases, the manufacturing process requires highly accurate control of the critical dimensions (CD) to produce sub-micron patterns. Variation in the critical dimension must be controlled to within less than 1/10 of the feature size in order to avoid changes in the device characteristics which would give rise to loss of product. This means that the equipment used for determining the CD measurement should have an accuracy of better than 10 nm. In this range of pattern sizes, optical methods cannot be used and it is necessary to move to an electron beam method for inspection.

In these processes, particle contamination plays a dominant role in determining the production yield and keeping it under control is a major concern. Generally, particle sizes larger than 1/5th of the pattern feature size cause defects. Consequently measurement techniques which will monitor this are particularly important in helping to limit contamination. Since various thin film formation techniques are also methods of inspecting the thickness and characteristics of the resulting layers are required. These can usually be obtained using optical methods.

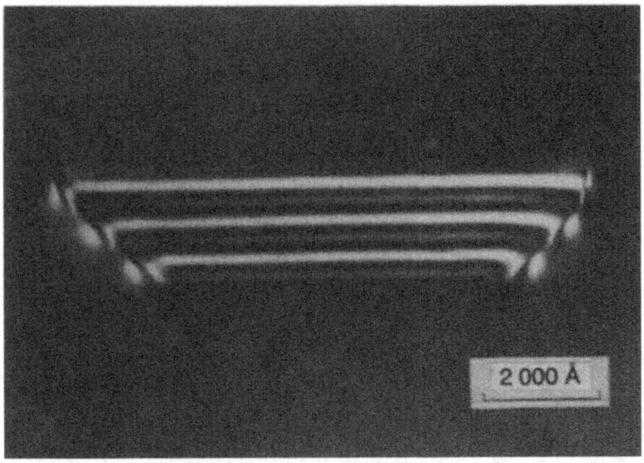

Figure 6.9 Transmission electron microscope image of the process induced defect (OSF). Accelerating voltage is 200 kV.

All materials used in the VLSI manufacture must be of high purity, since even a small amount of impurity will drastically degrade the characteristics and reliability of fabricated devices. This means the production line must satisfy high standards of cleanliness if contamination is to be controlled. Where an impurity (dopant) is introduced deliberately in diffusion and ion-implantation, the total amount of dopant needs to be precisely controlled which means that a suitable method of chemical analysis is needed.

6.2.2 Electron microscope techniques

Transmission electron microscope (TEM)

The transmission electron microscope (TEM) has a very high resolving power of the order of 10^6. This gives a resolution of 0.1 nm in a lattice image. However, such a resolution is drastically reduced with a thick sample because the electron beam cannot pass through it. Consequently, TEM is commonly used for examining sections of carefully prepared thin samples having a thickness of the order of several tens of nanometres. In TEM, some skill is necessary in interpreting the images obtained, since they can be formed by a number of mechanisms such as scattering, absorption, diffraction and phase contrast.

The main application of TEM in VLSI manufacture is looking at the crystallographic features and process-induced defects (offline) in order to help in fault diagnosis. Figure 6.9 shows a photograph of an oxidation-induced stacking fault (OSF) using TEM. OSF is a typical form of stack-

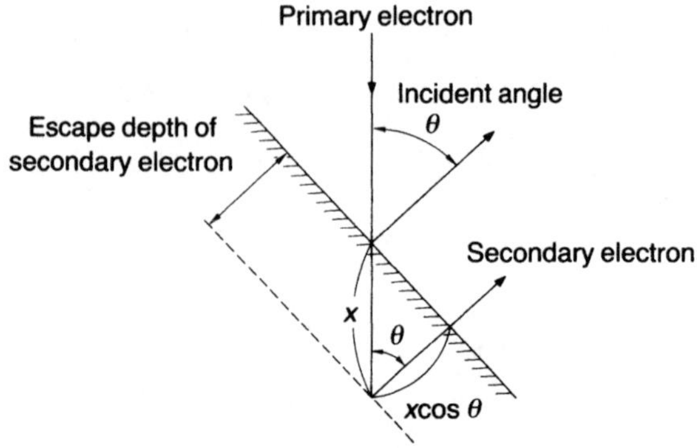

Figure 6.10 Variation of secondary electron emission yield with the effect of inclination of the object surface with respect to the incident electron beam.

ing fault induced by oxidation. As the silicon crystal lattice sinks uniformly to a given depth in the rectangular region of the photograph, it causes diffraction of the electron beams and a contrast in the image can then be clearly observed because of the local variations in diffraction intensity.

In addition to the observability of the crystallographic features and atomic level defects, an advanced feature of TEM is that it can be linked to optical equipment such as EDS (energy dispersive X-ray spectroscopy) and EELS (electron energy loss spectroscopy). This provides an analytical capability (analytical electron microscope) which is becoming increasingly important.

Scanning electron microscope (SEM)

In a SEM, a finely focused electron beam is used to scan the surface of the specimen in two dimensions. An image is formed by serially detecting the secondary electron signals emitted from the surface of the specimen. The resolution of the image is determined basically by the diameter of the electron beam, which means that a resolution of several nanometres is possible. The contrast in the image originates mostly from variations in the extent of secondary electron emission from the topographic features of the surface. The extent of secondary electron emission is influenced by

- surface topography,
- elements in the surface of the specimen or the work function (surface condition),
- the crystal lattice orientation and channelling effect, and
- electric or magnetic field at the surface of the specimen.

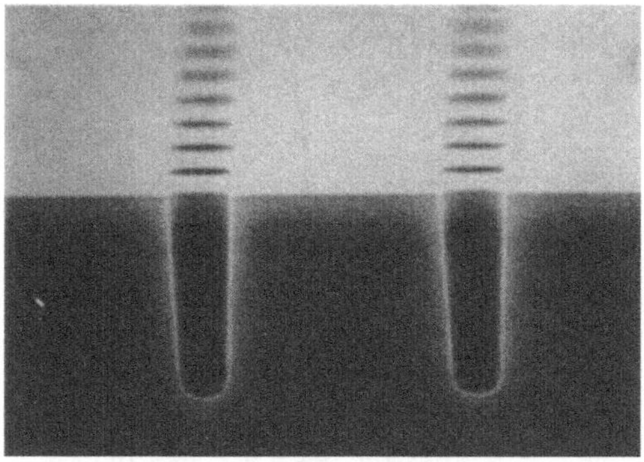

Figure 6.11 Cross-sectional view of the trench pattern by the cross-sectional SEM observation method.

In most applications, the surface topography plays the dominant role. The image contrast originating from the surface topography arises from the secondary electron emission which is proportional to $1/\cos\theta$, as can be seen from Fig. 6.10. This depends on the depth of penetration, (50 Å–100 Å) of the primary electrons to the point where secondary electrons are emitted. As the angle of incidence increases, $x\cos\theta$ becomes larger and there is an increase in secondary electron emission. At the edge, the secondary electron emission becomes large because of the increased penetration of the incident electrons in the specimen.

The major use of SEM is to control the processing conditions by observing a cross-sectional view of the wafer (cross-sectional method for SEM). Figure 6.11 shows a photograph of deep trenches formed by etching the silicon substrate obtained using this method.

6.2.3 Critical dimension measurement

Optical methods

Principle of the technique

In CD measurement, edge determination, which fixes the point of measurement, is particularly important. For the simplest case of a planar mask pattern and the ideal contrast a schematic illustration of the optical intensity profile obtained from an optical microscope is shown in Fig. 6.12 [60]. Although some spreading of the profile occurs near the edge due to diffraction, it can be determined by the intersection of the profile and some predetermined threshold level. In any particular case, the threshold level has to be adjusted depending on the measurement being taken because of complicated phenomena arising from variations in the thickness and the shape at the edge of the pattern.

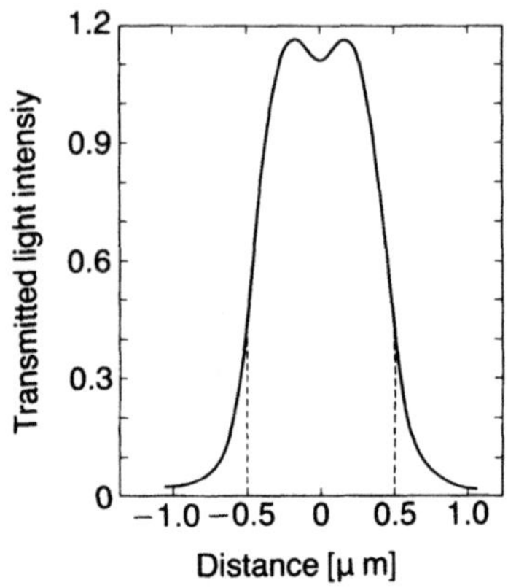

Figure 6.12 Schematic illustration of the optical signal intensity profile from an ideal mask pattern obtained by the image formation system of the optical microscope.

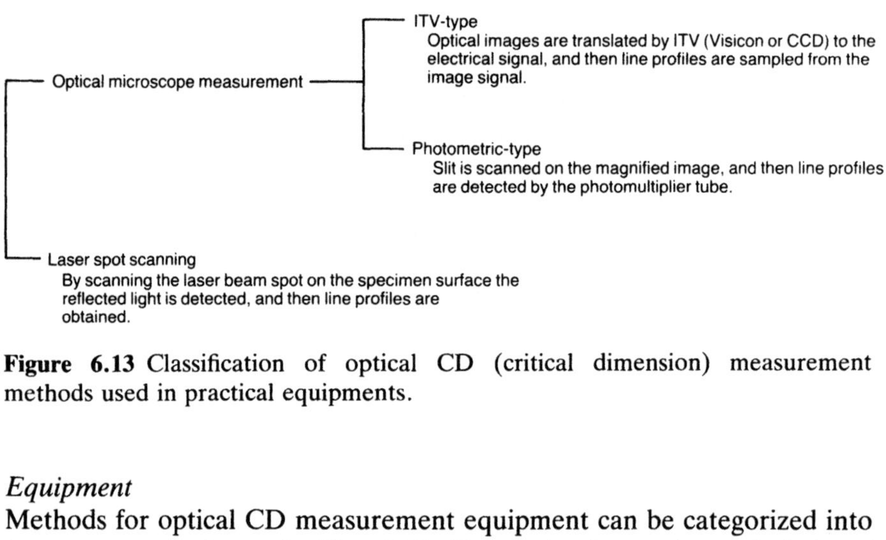

Figure 6.13 Classification of optical CD (critical dimension) measurement methods used in practical equipments.

Equipment

Methods for optical CD measurement equipment can be categorized into two groups, as shown in Fig. 6.13. One is to detect the optical signal intensity profile of the magnified image of the reflected light using an optical microscope. The other is based on detection of the intensity profile of the reflected light by scanning the focused laser beam spot [61]. The laser beam spot scanning method gives high resolution since the small spot is near the diffraction limit for the optical system and is obtained by using a bright laser optical source.

The laser spot is scanned by a mirror and the distance moved is measured, using a Michelson interferometer, to an accuracy of 0.01 μm. The reflected light from the specimen surface and the scattered light

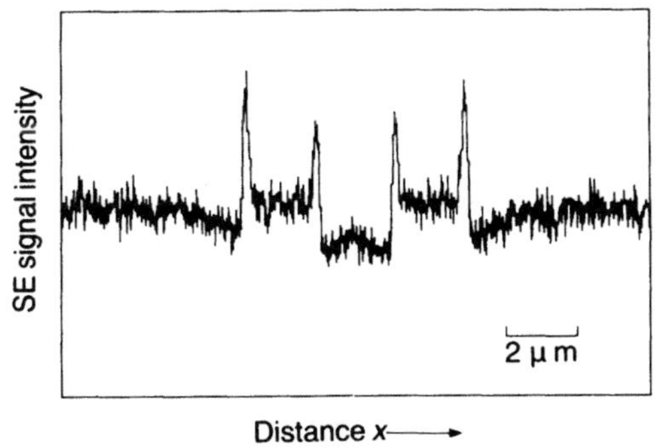

Figure 6.14 Measured SE signal intensity profile from two neighbouring photoresist patterns.

from the pattern edge are detected synchronously using a scanning laser beam. Equipment having a high spatial resolution with a beam spot of size 0.4 μm from a He-Cd laser (wavelength 325 nm) is now available commercially.

With optical equipment, the optical signal intensity profile changes with the lighting condition of the object being measured, so it is necessary to calibrate it beforehand using a more accurate instrument, such as an SEM.

Electron beam methods

Basic techniques

Far higher resolution is possible using a scanning electron microscope (SEM) than with light. The secondary electrons generated from the specimen surface exhibit sensitivity to the topography of the surface. The resolving power of an SEM is essentially dictated by the electron beam diameter, which means that the resolution can be 10 nm or better. Thus, by scanning the surface with a finely focused electron beam and detecting the secondary electrons, it is possible to obtain measurements with very high absolute accuracy.

Figure 6.14 shows the secondary electron (SE) signal intensity profile from two neighbouring photoresist patterns. The features and the method used for generation of the SE signal intensity profile can be explained from an analysis of the topographic contrast signal [62]. The SE signal intensity decreases near the pattern being monitored due to the collision of secondary electrons in the slope region. This results in a decrease in the secondary electrons reaching the SE detector. This phenomenon is highlighted in the region between two neighbouring patterns. A minimum value of the SE signal intensity appears at a position on the outer boundary because only half of the beam size is seen. It then increases linearly

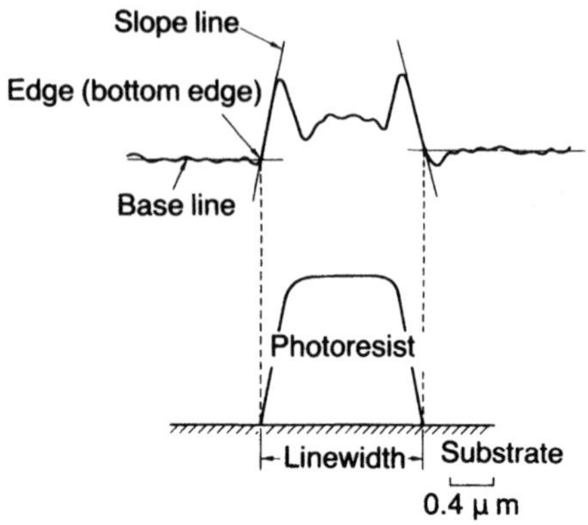

Figure 6.15 SE signal intensity profile from the photoresist pattern on the silicon substrate and a principle of the linear regression method.

over the slope region of the pattern. This is caused by changes in the SE emission yield, corresponding to a large change in the cross-sectional area at the bottom edge.

Near the top, the change in the SE signal intensity profile is small because of the difference in the detection efficiency of the secondary electrons emitted from the slope region. The maximum value of SE signal intensity appears where the beam crosses the boundary and exposes only half the spot diameter.

Automatic techniques

An automatic measurement procedure known as the 'linear regression method' utilizes the high spatial resolution of the electron beam and guarantees accuracy, even when the materials and the slope angle of the pattern edge vary [63]. In this approach, the SE signal intensity profile near the edge is approximated by two lines. One is the average line of the SE signal intensity profile on the substrate, and the other is the slope line corresponding to the slope of the pattern edge, as shown in Fig. 6.15. The cross-over point of the two lines corresponds to the bottom edge. The distance between the two cross-over points is defined as the line-width. By approximating a portion of the SE signal intensity profile by a straight line (analogous to integration), the influence of charging-up and noise are averaged and eliminated, resulting in better reproducibility. With this method, an absolute accuracy of 9 nm is possible for slope angles in the range of 72° to 85° of the photoresist patterns.

Equipment

Electron beam CD measurement equipment which satisfies the requirements of on-line process measurement is now available for VLSI manufacture. Since wafers have to be returned to the manufacturing line after taking measurements, the equipment must satisfy the following requirements, in contrast to that of conventional SEM.

1. The test must be non-destructive, i.e. there must be no radiation damage of the device characteristics during the measurement and no effect on the following stages. In addition, the measurement should not require special sample preparation. This means a low voltage, i.e. 1 kV, in order to avoid radiation damage and charging-up problems of the insulating materials such as the photoresist. However, to overcome the deterioration in image quality from the low acceleration, improvements in the electron gun and lenses are necessary.
2. It must be capable of handling of large size wafers automatically at high throughput and should be particle-free. This is possible with cassette to cassette wafer conveying, automatic stage movement and computer control. The operation of the electro-optical system (EOS) needs great skill in operation and necessitates automatic control of the SEM.

6.2.4 Determination of particle concentration

Principle of measurement

The method is based on illuminating the wafer so that light scattered by the particles can be detected. It is convenient to start by considering light scattering by spherical particles suspended in air [64]. In this case, the characteristics of the scattering cross-section depend on the diameter of the particle and wavelength of the incident light. The scattering cross-section, S, can be obtained from the following formulae:

$$\begin{aligned} &d \gg \lambda \quad S = 2\pi\,(d/2)^2 \\ &d \sim \lambda \quad S = k\pi\,(d/2)^2 \quad \text{(Mie scattering)} \\ &d \ll \lambda \quad S \propto 1/\lambda^4 \qquad \text{(Raleigh scattering)} \end{aligned} \tag{6.1}$$

where d is the particle diameter, λ is the wavelength of the incident light and k is the Mie scattering coefficient.

For particle diameters less than 1 μm (comparable to the wavelength of incident light), the scattering is given by the Mie formula. The cross-section, S, of a polystyrene latex sphere (refractive index $n = 1.56$) for Mie scattering is shown in Fig. 6.16 for an incident light wavelength of 780 nm.

In light scattering by spherical particles on a wafer, it can be assumed that the beam is subsequently reflected by the surface and enters the detector (Fig. 6.17). The scattered light intensity T_1 for spherical particles on a surface can be approximated by:

$$T_1 = T_B + RI_F \approx RI_F \tag{6.2}$$

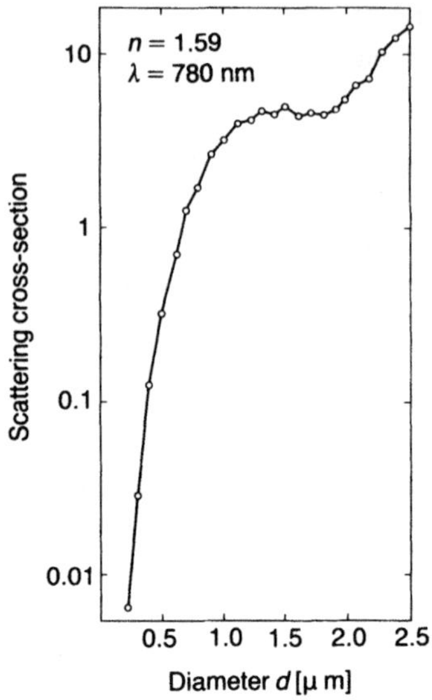

Diameter d	Size factor α	Scattering coefficient K	Scattering cross-section S
0.2	0.18	0.2	0.0062
0.5	2.01	1.7	0.333
1.0	4.03	4.05	3.20
1.5	6.04	2.9	5.31
2.0	8.05	1.8	5.56
2.5	10.01	2.85	14.0

Figure 6.16 Relationship between the scattering cross-section and the diameter of spherical particle in the Mie scattering.

Figure 6.17 Light scattering from a spherical particle on a wafer.

where R is the specular reflectance of the surface. It is assumed that the forward scatter contribution, scaled by the coefficient, R, is the largest for Mie scattering.

Wafers must be monitored on-line after thin film deposition. To achieve the same sensitivity, independent of the underlying material, it is necessary to adjust R for the material using (6.2). This generates almost the same characteristic curve in all cases, as can be seen in Fig. 6.18.

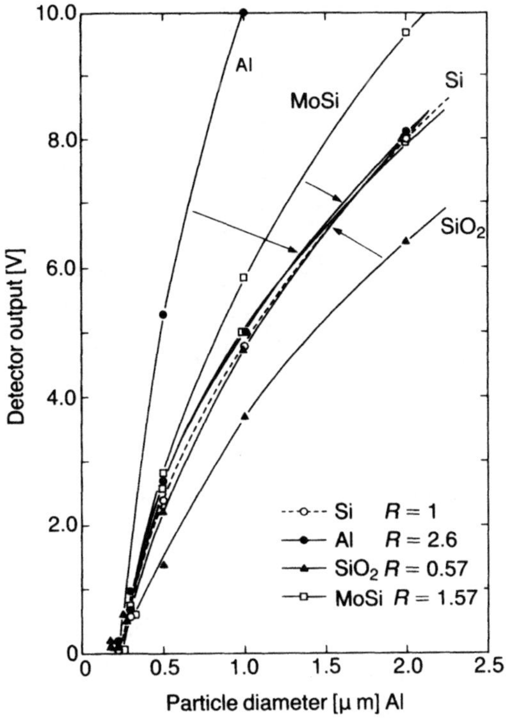

Figure 6.18 Light scattering characteristics of spherical particles on various underlying substrate materials and the compensation by the relative reflectivity (*R* coefficient).

Particle detection equipment [65]

The total particle count, shape of particle, and particle size are obtained from light scattering measurements after suitable data processing. This can be done on-line by automating the handling of the wafer samples. Laser light is well suited for this duty since it is a bright optical source. A beam diameter of 10–100 μm is formed by the optical lens system which then scans the wafer surface. Two scanning methods are commonly used: one is an X-Y raster scanner employing a combination of a scanning or polygon mirror and shifting of the wafer stage, the other is a spiral method using a combination of wafer rotation and shifting. The light scattered from particles is collected by an optical fibre, elliptical collection mirror or an integrating sphere and converted to an electrical signal by a photomultiplier.

Figure 6.19 shows a typical example of the detector configuration (TOPCON; WM-3). A laser diode (λ = 780 nm) is used as the optical

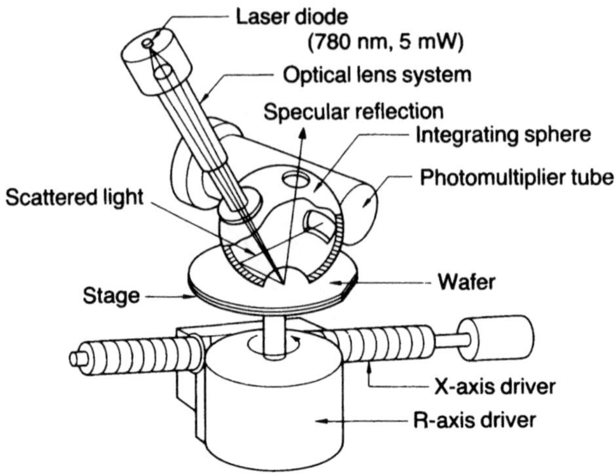

Figure 6.19 Schematic diagram of practical detection system (TOPCON; WM-3).

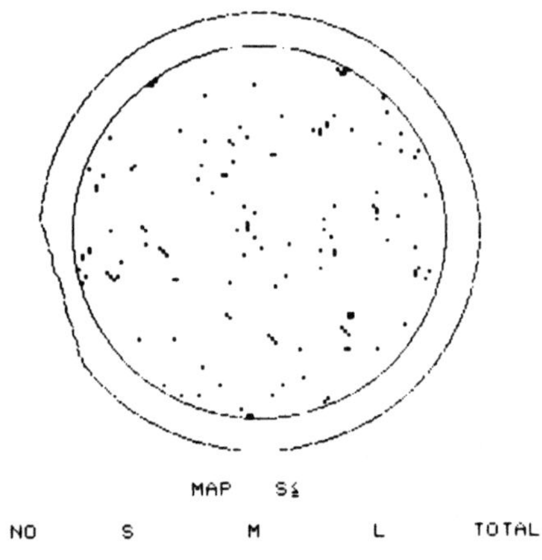

Figure 6.20 Typical example of measured result of the particle contamination on silicon dioxide grown wafer.

source, and the beam spot is focused to 50 μm. Scattered light is collected by an integrating sphere. As the laser beam scans, the wafer mounting is rotated and shifted by stepping motors, resulting in spiral scanning over the wafer surface. The scattered light is collected and analysed, and the peak height of the pulse signal corresponds to the particle size.

By detecting the location of the scattered light and the peak height of the signal, the particle size, shape of particles and the total particle count can be determined. Figure 6.20 shows a typical set of results.

6.2.5 Film thickness measurement [66]

Ellipsometer

It is possible to measure the thickness and the refractive index of a thin film using the principle of ellipsometry. This method is capable of high accuracy and can be used for ultra thin films of the order of a few monolayers. It is used only for gate oxide thickness measurements of several hundred ångströms. The film quality inspections are based on refractive index and can be used during normal operation even though the operational procedure and calculations which it entails are rather complex.

Ellipsometry involves changes in the state of polarized light following reflection of a precisely angled beam of known wavelength from a surface. Two factors are involved: the relative phase difference Δ between two components of polarization (P or S) and the azimuth Ψ, i.e. the ratio of the reflected amplitudes of the P and S components. The ellipsometer measures Δ and Ψ from which the thickness, d, and refractive index, n_2, can be calculated.

The values of Δ and Ψ are related to optical constants such as wavelength, λ, angle of incidence, ϕ, refractive index of the atmosphere, n_1, refractive index of the thin film, n_2, refractive index of the substrate, n_3, and film thickness, d. The procedure involves calculating the unknown parameters, d and n_2 from the measured values Δ and Ψ assuming λ, ϕ, n_1 and n_3 are known.

Figure 6.21 is a schematic illustration of the system configuration for an automatic ellipsometer. Linearly polarized light is reflected from the specimen and the ellipsoidally polarized light is measured by the analyser. Values of Δ and Ψ are obtained from the relationship between the angle of rotation of the analyser and the detected light intensity, which can then be used to calculate d and n_3. The measurement and calculation procedures are complex so it is convenient to automate the operation so as to give the wafer thickness and refractive index directly. An automatic ellipsometer has an accuracy of ± 3 Å when used for thermal oxide single layer thickness measurement. The thickness of each layer in the multilayer structure can be separately measured using light of different wavelengths.

Optical-interface method

The thickness of a thin film can also be measured using optical-interference. If white light is incident normally, the reflected light (spectrum) will show maxima for each wavelength in the form of an interference patterns from which the film thickness can be calculated, given the refractive index.

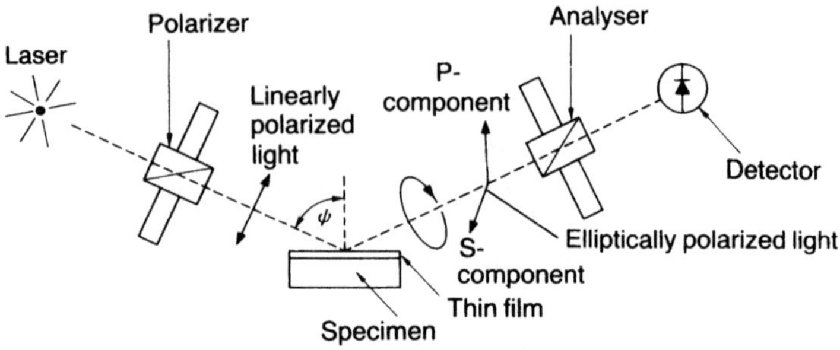

Figure 6.21 Schematic diagram of an automatic ellipsometer.

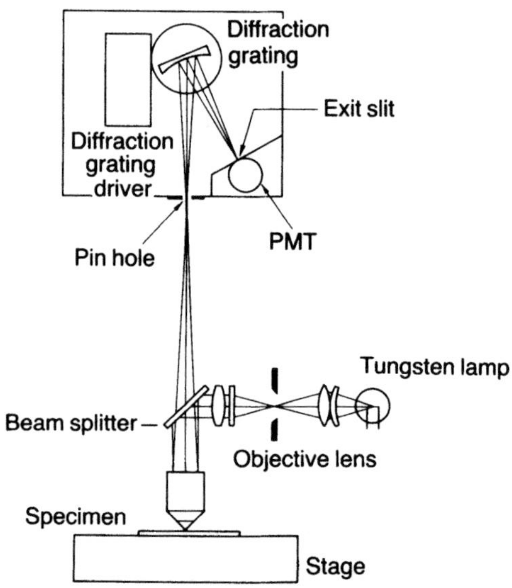

Figure 6.22 Schematic diagram of optical system in the practical equipment using the optical-interference method.

Figure 6.22 shows a schematic diagram of a typical optical system based on the spectroscopic method (Nanospec AFT). White light from a tungsten lamp is incident on to the film. The reflected light is passed to the optical grating which then photomultiplies the signal for spectral analysis. This technique has a reproducibility better than ±5% for a thickness of 100 Å.

6.2.6 Impurity analysis

Chemical methods are used mainly for determination of elements. Auger electron spectroscopy (AES) and secondary ion mass spectroscopy

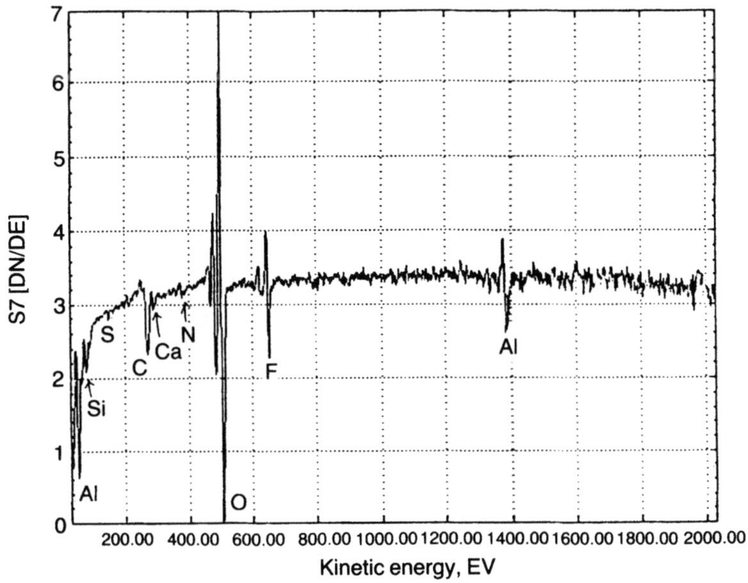

Figure 6.23 Auger spectra from aluminium surface showing some contaminants.

(SIMS) are particularly useful and are located near the production line [67]. AES analyses Auger electrons ejected from an atom after core-level ionization by electron beam bombardment of the specimen surface. The escape depth of Auger electrons is generally around 20 Å. This makes it possible to analyse the solid surface regions. Detection limits are in the range 10^{19}–10^{20} atoms cm^{-3} and the difference in detection sensitivity for various species is small in comparison with other surface analysis techniques. For in-depth analysis, the sample is ion-milled by noble gas ions such as Ar^+. Data are obtained by interrupting the milling process at regular intervals and obtaining the Auger spectra. A two-dimensional analysis of electron beam scanning of the surface and the Auger electron image (scanning Auger microscope image) is the most popular method. Analysis of a small region can be made using a finely focused primary beam (Micro-AES). Figure 6.23 shows Auger spectra from an aluminium surface, indicating the effect produced by contaminants.

In the SIMS method, an ion beam sputters material off the surface of a sample. This results in secondary ions which are then detected using a mass spectrometer [68]. Noble gas ions such as Ar^+ are commonly used as primary ions. Positive caesium ions beams are generally useful for producing high negative ion yields of the electronegative species and O_2^+

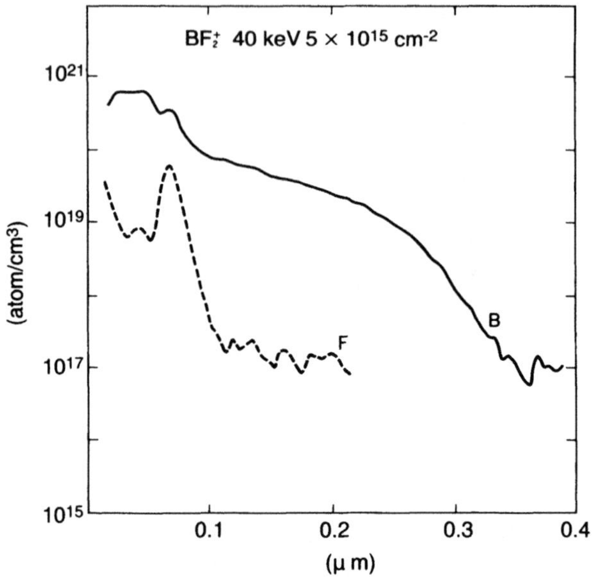

Figure 6.24 Concentration profiles of B and F in Si by the SIMS method.

ion beams are usually used for generating high positive ion yields from electropositive species.

In-depth profiles are easily obtained using the sputtering effect of a primary ion beam. The vertical (depth) resolution is better than 1 nm. However, the escape depth of sputtered secondary ions is a few monolayers so the actual vertical resolution generally obtained is a few tens of nanometres because of the influence of the surface micro topography. A major feature of the SIMS method is the high sensitivity compared with other surface methods. Detection sensitivities of better than 10^{16} atoms cm^{-3} are achievable for boron in a silicon matrix. The difference in detection sensitivity for various species runs into four figures, which is significantly greater than for any of the other analytical methods. Figure 6.24 is an example of a SIMS analysis showing the concentration profile of boron implanted into silicon. Profiles of B and F after the annealing to activate implanted ions are shown. The miniaturization required by VLSI means that monitoring and control of the concentration profiles of implanted impurities using SIMS method will become more important.

7

Production of silicon wafers and environmental problems

As can be seen from the flow diagram of the case study, a variety of materials are used in the fabrication of semiconductors, including silicon wafers, chemicals, gases and water. These raw materials are produced mostly by the chemical industry. This chapter considers the market for, and production of, these materials. A crucial factor in the operation is the control of the environment, particularly dust and waste from the process, both of which will have a direct effect on the yield of the final products. Unreacted chemicals and gases from the process cause pollution if they are not treated properly and so it is necessary to consider how they need to be handled to avoid such problems.

7.1 SILICON WAFERS

7.1.1 Special market features

The process of producing silicon wafers can be summarized quite briefly as the reduction of silica (SiO_2) to form silicon metal followed by purification and growth of high purity polysilicon crystals. The last stage involves pulling single rods of crystal from a melt of polysilicon, after which they are sliced and polished to form the wafers. The production of 1 kg of polysilicon requires 24 kg of silica and consumes 580 kW h of electricity. The size of the wafers has grown each year. In 1990, wafers of 10 inches (25 cm) diameter were being pulled by the Czochralski method: with the floating zone melting, the size is limited to 8 inches (20 cm). The scope for using 8 inch wafers is declining and the pressure is to try to increase the wafer size.

Dopants such as boron or phosphorous, which are added to provide the desired electrical properties, need to be distributed uniformly along the axis of the crystal. To ensure this, the top and bottom of the ingot are cut off, so only that part of the ingot having the desired concentration is used. The next stage involves grinding the side of the ingot to give the appropriate wafer size. This results in about 50% of the crystal being wasted. A further 50% of the remainder is lost during the slicing and polishing steps which produce wafers having a thickness of 400–600 μm. Consequently, only about 25% of the original single crystal of silicon is effectively used. This is likely to be even less when the losses resulting from cutting the wafers to obtain the final chips are taken into account. Based on metallic silicon, as distinct from the single crystal, the yield is therefore very low.

7.1.2 Production processes

Polysilicon

When high purity silica is reduced in an electric arc with charcoal and/or coke, silicon metal of 98% purity is obtained. The overall reaction can be written as:

$$SiC(solid) + SiO_2(solid) = Si(solid) + SiO(gas) + CO(gas)$$

In order to remove the impurities present in the silicon, such as Al, Fe and Cr, purification is carried out by crushing the metal and converting it into trichlorosilane in hydrogen chloride vapour:

$$Si(solid) + 3HCl(gas) = SiHCl_3(gas) + H_2(gas)$$

The reaction is carried out in a catalytic fluidized bed reactor at 300°C. After distilling off the trichlorosilane, it is mixed with hydrogen and sent to a bell type reactor to precipitate the polysilicon on a silicon rod using chemical vapour deposition (CVD):

$$2SiHCl_3(gas) + 3H_2(gas) = 2Si(solid) + 6HCl(gas)$$

Polysilicon crystal rods of 20 cm diameter and 2–3 m in length can be obtained with the CVD process operating at 1100°C for 200–300 hours in what is known as the Siemens process (Fig. 7.1). The impurity content in the polysilicon must be less than 2 p.p.m. for carbon and less than the 1 p.p.b. level for the group III or V elements, which act as dopants in the single crystals of silicon.

An alternative method of producing polysilicon crystal is the monosilane process in which mixtures of powered metallic silicon and magnesium are heated to 500°C in a hydrogen atmosphere to form magnesium silicide (Mg_2Si) which is allowed to react with ammonium chloride in liquefied ammonia below 0°C to produce monosilane (SiH_4, boiling point −112°C).

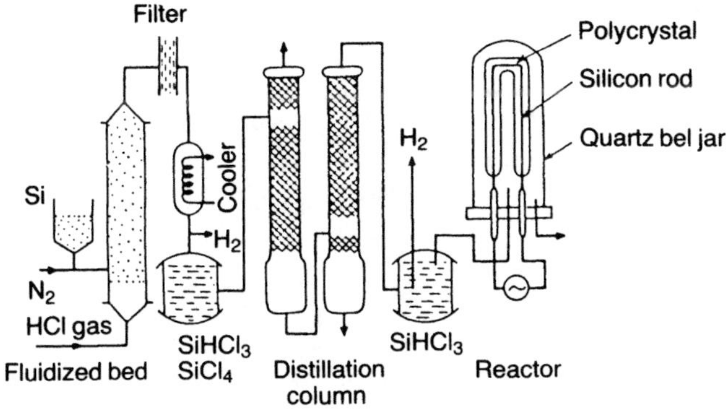

Figure 7.1 Process for production of silicon polycrystal.

After purification, the monosilane is thermally decomposed at 800–900°C and the polysilicon is deposited on a silicon rod substrate.

Of the two processes, the the Siemens is most widely used worldwide.

Single silicon crystals

Rods of single crystal are produced from polysilicon either by the floating zone melting (FZ) or Czochralski (CZ) method. Although the FZ has the advantage that oxygen or metal inclusions in the single crystal can be avoided because no crucible or container is used in the crystal growth, it is difficult to produce large ignots and to ensure uniform distribution of the dopants in the crystal. Consequently, this process is not used for making VLSI wafers.

In the CZ method, polysilicon crystals are melted in a crucible and a seed crystal is attached to the tip of the pulling rod which is then withdrawn while at the same time continuously rotating it. A schematic drawing of the equipment used for crystals of 8 inches (20 cm) diameter and 1.65 m in length is shown in Fig. 7.2. To produce dislocation-free single crystals, the diameter of an ingot is first reduced to about 3 mm and then allowed to grow to 3 cm, during which time the existing dislocations are drawn to the surface and disappear.

VLSI wafers contain small amounts of elements belonging to either groups III or V which are added to provide the specific electric properties. Wafers can be either positive (p) or negative (n) type. For CMOS wafers, p-type materials are used and contain about 20 p.p.b. of boron uniformly distributed. With the CZ method, the dopant is added to the melt and the single crystal pulled from it contains dopant with the concentration determined by the equilibrium distribution (partition) coefficient. Since this is not unity, the concentration of the dopant in the crystal is different from

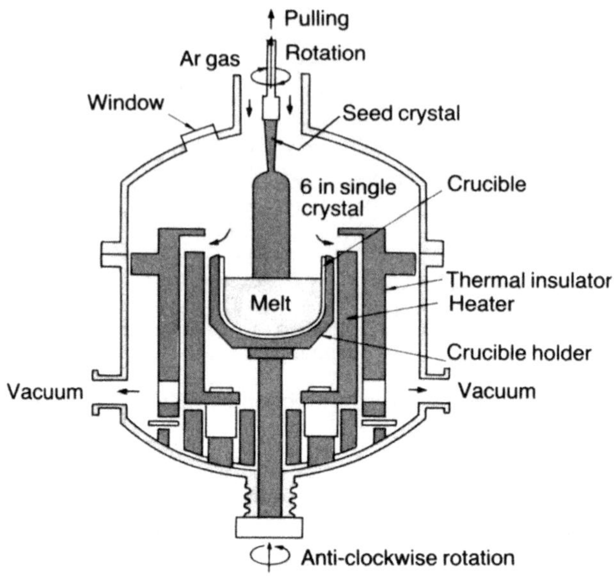

Figure 7.2 Outline of puller of 6 in. single crystals.

that of the melt. Consequently, there is segregation of the dopant which gives rise to a concentration distribution in the ingot, mainly in the axial direction. As a result the concentration in the rod is not uniform, which explains why the ends of the ingot are cut off, as noted above.

In addition, non-uniformity of the dopant occurs in the axial and radial directions arising from the convective flow of the melt due to the temperature distribution and rotation of the rod and crucible. In order to reduce the convective flow, a magnetic field can be applied to the melt. The pull rate not only influences the impurities and the formation of dislocations in the crystal but is also a key factor in determining the production rate. The diameter of the single crystal is primarily determined by the pull rate and thermal conditions. Overall the diameter is inversely proportional to the pull rate.

7.1.3 Wafer shaping

In order to obtain wafers from single crystal rods, some cutting and polishing to achieve a mirror-like surface is required. The sequence of the processes is illustrated in Fig. 7.3. The stages are described below.

1. *Cutting* of the ingot into blocks using a belt or disk saw with diamond blades.

Processing	Outline
(1) Cutting and inspection	
(2) Surface grinding and shaping	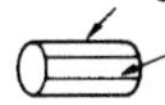
(3) Slicing	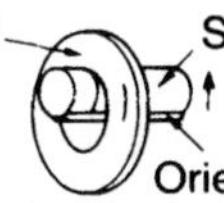
(4) Lapping	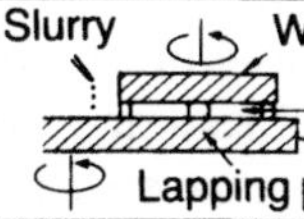
(5) Edge contouring	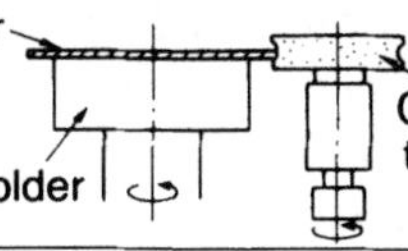
(6) Etching	
(7) Gettering	
(8) Polishing	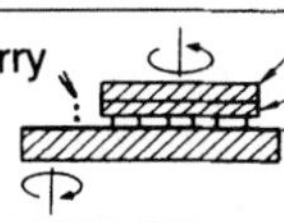
(9) Cleaning	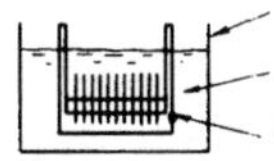
(10) Inspection	
(11) Packaging	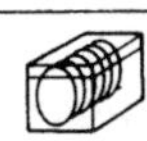

Figure 7.3 Production process of mirror-polished surface wafers (modified from [71]).

2. *Surface grinding and orientation.* Surface grinding is necessary to bring the ingot to the desired diameter because it needs to be slightly oversized to compensate for the difficulty in obtaining exactly the desired diameter during growth. The orientation of the ingot is deter-

mined by X-rays and an orientation flat (OF) is ground on to the surface to mark the crystallographic orientation. The OF is later used to align the wafer and masks correctly.

3. *Slicing the ingot to form wafers*. The ingot is fixed on to a mount and cut using a diamond blade fitted to the inner rim of a circular saw having a thickness of 270–350 microns. The actual thickness of the wafer depends on the diameter of the ingot, but is typically about 675 ± 15 microns for a 6 inch wafer: the warp is 20 microns.
4. *Lapping the sides*. Wafers are lapped mechanically, using alumina or SiC fine particles to remove distortions induced by the cutting operations and to make them flat.
5. *Bevelling of edges*. The wafers are edge-contoured using a diamond grinder so as to avoid breakage during subsequent LSI fabrication.
6. *Etching*. In order to remove distortions induced by the lapping processes, each wafer surface is etched in acid or basic solutions, to a thickness of about 10 μm.
7. *Extrinsic gettering*. Gettering is used to capture impurity elements such as oxygen. This is done by 'sandblasting' the reverse side of the wafer generating internal defects in the bulk of the wafer which effectively incapacitate the undesirable elements.
8. *Mirror polishing*. The wafer is fixed in a holder with wax and rotated relative to a polishing pad. A polishing slurry containing colloidal silica is fed on to the surface of the wafer and polishing proceeds chemically as well as mechanically until a good surface finish is obtained. Typically, the surface of a 6 inch (15 cm) wafer should be flat to within 3 μm.
9. *Washing*. A variety of processes is used including organic solvents, acid solution in conjunction with mechanical action involving brushes and high pressure water to remove the organics, metal contaminations and micro particles from the surface.
10. *Inspection*. Before shipping, an inspection is carried out to check the electrical properties, dimensions and cleanliness of the surface using incident light.
11. *Packaging*. The wafers are packed in a special carrier box which is sealed under dry nitrogen or atmospheric air.

7.2 CHEMICALS USED IN VLSI PROCESSING

Many different chemicals are used in semiconductor manufacture, including photoresist resins, gases, chemical reagents and ultra-pure water. They are considered below with respect to their use, properties, production processes, market and related issues.

7.2.1 Resins [69–73]

As mentioned in Chapter 5, lithography is still used so photoresist resins play an important role. The mask is a photosensitive polymer resin and transfers the pattern on to a wafer surface using the photoreactivity of the resin. Many resins have become available since Kodak succeeded in producing the necessary synthetic chemicals.

Basically, UV light (mainly λ = 436 nm) is used to form a pattern on the wafer and an electron beam is then used to produce a quartz plate. Discussion here is restricted to resist resins activated by UV rays.

The resist solution consists of a resin and organic solvent (e.g. xylene) which is spread over the wafer to give a uniform layer of thickness of about 1 μm on a disk rotating at 2000–5000 rev/min. After baking at 80°C to remove the solvent, the mask pattern is transferred to the wafer by irradiating it with UV light. The exposed pattern is developed using an alkaline or organic solvent followed by etching. Two types of resist resins are available. One is negative and makes the exposed sections insoluble in the developer by forming a cross-linked polymer. The other is a positive resist which results in a soluble exposed section where a photodecomposition reaction takes place.

In general the desirable properties of photoresist resins are that they have superior photosensitivity and resolution and have a strong resistance to etchants such as aqueous HF solution or etching gases. These properties are important in order to ensure a precise image of the desired pattern.

Until the production of the 64 Kbit DRAM, ring-forming rubber resists of negative type were almost always used. Nowadays positive resists of naphthoquinonazido are used for 256 Kbit DRAM fabrication. The ring-forming rubber resists are blended polyisoprene with bisazide as the photosensitive component: UV irradiation generates reactive nitrene radicals ($R{-}\ddot{N}{=}\ddot{N}{=}\ddot{N}{:} \xrightarrow{h\upsilon} R{-}\ddot{N}. + N_2$), and a hardened polymer film is produced by the photo-crosslinking. The resin on the unexposed section is stripped off by the developer liquid (typically organic solvents) leaving a thin film of resist on the wafer.

An alternative method is based on a positive resist such as a mixture of Novalack-phenol resin (MW = 5000–20 000) and naphthoquinonazido. This decomposes photochemically to produce N_2 gas and ketenes by a transformation reaction finally yielding indene carboxylic acid (see section 5.2.5). The carboxylic acids are soluble in alkaline solution while the unexposed section has a high resistance to the solution because it is stabilized by the naphthoquinonazido. The irradiated resin is then developed using an alkaline solution such as $(CH_3)_4NOH$ and a thin film of positive resin is left. Aromatic compounds have a high resistance to dry etching so the above resist consists of both an aromatic resin and a photosensitive agent.

Ring-forming rubber resists are attractive because of their good properties: firm adhesion, high chemical resistance and low cost. At the present time, they tend not to be used for the 1 Mbit DRAM production because they swell during development and do not satisfy the minimum feature size for 1 Mbit DRAM (1 μm). Novalack resin resists give acceptable resolution because they do not swell and have a higher resistance to dry etching and so have been widely used. There has been a good deal of work to find an alternative negative resist which has the low swelling characteristics found with polymethyl-isopropenyl ketone and polyvinylphenol. There is also a good deal of activity in searching for a new resist suitable for use with X-rays or to enable high resolution micro-lithography to be developed.

The mask is produced by exposure to an electron beam (EB). The EB can be focused to within 5 Å and has a high enough energy to form a pattern on the resist directly by cutting the cross-linking of the main chain of the polymer resin. However, EB lithography is limited to mask-making because it is not possible to use bundles of multiple beams. Also, EB tends to be dispersed in the resist and even causes inaccurate pattern drawing. Because of this, a very thin resist film has to be used. Positive resists of the polymethacrylic acid resin and polymethylisopro-phenyl ketone type are used, and silicon-containing resist polymers (both positive and negative types) are also used for multi-level lithography to prevent EB dispersion. To limit the loss of resolution arising from ray-reflection for the wafer and the interference of rays in the resist, CEL (contrast enhanced lithography) is used. This enhances the resolution by employing a resist coated with substances transparent to the rays which makes the masking of the unexposed section more effective.

Typically, the annual market for photoresists is of the order of 10–20 billion yen in Japan. Positive resists take about 50% of the total production and more than 60% of the sales. As the market grows, many companies are starting to develop new resists for micro-lithography using EB and X-rays. Because large numbers of resists are used in various stages of VLSI manufacturing, there is pressure to reduce both the cost and number used during processing.

7.2.2 Gases

More than 40 kinds of gases are used in VLSI processes and are summarized in Table 7.1.

Large volumes of nitrogen and hydrogen are consumed in purging and as carrier-gases (N_2 6 million m^3/month, H_2 3 million m^3/month in Japan). Argon and helium are also used. The gas flow in all the piping needs to be continuous to prevent contamination with other gases such as oxygen,

Table 7.1 Gases for semiconductor production

Process	*Gas*
Thermal oxidation	HCl, Cl_2, O_2, H_2, Ar
Epitaxy	SiH_4, SiH_2Cl_2, $SiCl_4$, HCl, PH_3, AsH_3, B_2H_6, $SiHCl_3$
Diffusion	PH_3, AsH_3, B_2H_6, BCl_3, O_2, H_2, N_2, Ar, $POCl_5$
CVD	
Oxide film (Oxidized)	SiH_4, SiH_2Cl_2, PH_3, AsH_3, B_2H_6, O_2, N_2, N_2O, CO_2, H_2
Polysilicon film	SiH_4, SiH_2Cl_2, N_2, H_2
Nitride film	SiH_4, SiH_2Cl_2, NH_3, N_2, Ar
Metal film	WF_5, H_2, SiH_4, $MoCl_5$, $Al(C_4H_9)_3$
Sputtering	N_2, Ar
Ion-implantation	PH_3, AsH_3, BF_3, H_2, N_2
Dry (plasma) etching	CF_4, $CClF_3$, CHF_3, C_2F_6, C_2ClF_5, CCl_4, Cl_2, NF_3, SF_6, BCl_3, O_2, N_2, He, Ar, HCl
Carrier (purging)	N_2, H_2, Ar, He, CO_2

carbon dioxide, carbon monoxide or water. In particular, 2 million m^3/month of argon are used to provide an inert atmosphere in a typical single crystal silicon production process.

The gases used for epitaxy, diffusion, CVD, sputtering, ion-implantation and etching are known as materials gases in semiconductor plants (Table 7.1). They are typically hydrides or halides containing group III or V elements. The hydrides are the most suitable since they are highly reactive and the decomposition products do not have a deleterious effect on the process. In dry-etching, halogenated gases are mainly used and this produces volatile compounds such as $SiCl_4$. It is easier to control the speed of etching and the anistropy with halogenated gases rather than strongly reactive gases such as Cl_2 or F_2. The annual consumption of gases used in semiconductor manufacture in Japan is approximately 40 t of SiH_4, 30 t of SiH_2Cl_2, 140 t of CF_4 for dry-etching, 15 t of BCl_3 for etching aluminium, and 2.5 t of doping gases (AsH_3, PH_3 and B_2H_6) [74].

N_2 and Ar are produced by low temperature distillation and pressure swing adsorption (PSA) from plants located on site. The gases are produced by the (Japanese) chemical industry using a variety of processes and supplied in dilute form as ultra-pure carrier gases (i.e. at impurity concentrations of the p.p.m. level to 10%).

They should meet the following specifications:

1. impurities such as oxygen, water, hydrocarbons, carbon dioxide, carbon monoxide, and metal elements (Na, K, S and P) should be

Table 7.2 Chemical reagents for semiconductor production

Process	*Chemical reagent*
Washing	H_2SO_4, HNO_3, HCl, NH_4OH, H_2O_2, N_2H_4, Acetone, Trichloroethylene (Trichlene), Methanol, Isopropanol, Xylene, Toluene
Etching	H_2SO_4, HNO_3, HCl, HF, NH_4F, H_3PO_4, KOH, N_2H_4, $NH_2(CH)_2NH_4$, Acetic acid

below the levels of detection since they all have an adverse effect on the product; and
2. the concentration of the dilute gases should be kept constant at the prescribed levels. This demands sophisticated techniques for both purification and storage.

The purity of carrier gases should be at least 99.99% at the production site, which may be further refined on site using a catalytic processes followed by adsorption on zeolites or activated carbons until the purity reaches 99.9999999%.

Excluding the large volume carrier gases, the market for the semiconductor material gases is of the order to 10 billion yen per year in Japan. SiH_4 at 4 billion yen is the largest, with etching gases such as CF_4 and BCl_3 about 1.5 billion yen and doping gases such as AsH_3, PH_3 and B_2H_6 of the order to 0.5 billion yen.

Most of the semiconductor material gases are hazardous and flammable and the chemical, physical and hazardous properties are not well documented, largely because the demand for them has been small up to now. There is a large gap in knowing how to handle such gases in order to avoid a serious accident. Most of the other gases for fine micro-lithography have not been widely used and their properties are not generally known.

7.2.3 Reagents

Many kinds of reagents are used in various stages of the process, e.g. washing, etching and photo-lithography. Table 7.2 lists the chemical reagents used for washing and etching. They are typically pure organic liquids or aqueous acid or alkaline solutions.

Organic contaminants may be removed by washing with trichloroethylene and oxidizing with NH_4OH-H_2O_2 and fuming nitric acid. An HCl-H_2O_2 aqueous solution is used for removing ionic metal contaminants from the wafers. Lower alcohols and ketones can be used as solvents, instead of water, and for drying. In the wet etching process, HF solutions containing HNO_3, H_2SO_4, acetic acid and NH_4F are commonly used.

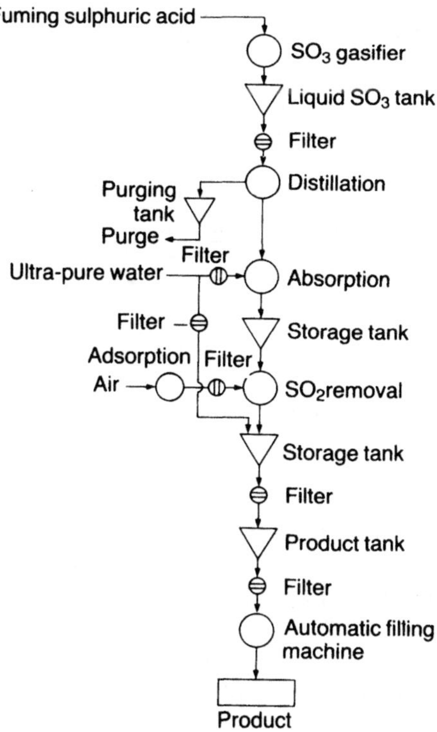

Figure 7.4 Flow sheet for production of sulphuric acid for semiconductor use [75].

There should be no impurities in the reagent nor any small dust particles or metals because these affect the production processes adversely. The metal impurities in commercially available reagents for semiconductor production need to be less than a p.p.b. Alkaline, alkaline earth and transition metals cause electrical damage, and doping elements such as B, As and P should be absent. The reagents listed in Table 7.2 are for chemical use and special care is needed to ensure they are metal-free and suitable for use in semiconductor manufacturing. To avoid contamination occurring during production, storage and transportation, all equipment should be made of glass or fluorinated resin rather than metals such as stainless steel.

Nitric, hydrochloric and sulphuric acids are purified using azeotropic distillation. To minimize the impurities, about 30% of the feed is lost as effluent. Industrial grade sulphuric acid contains many kinds of metal. Supplies to the semiconductor industry are made from metal-free sulphur derived from the hydrodesulphurization processes in petroleum refining. Figure 7.4 shows the process by which anhydrous sulphuric acid is distilled

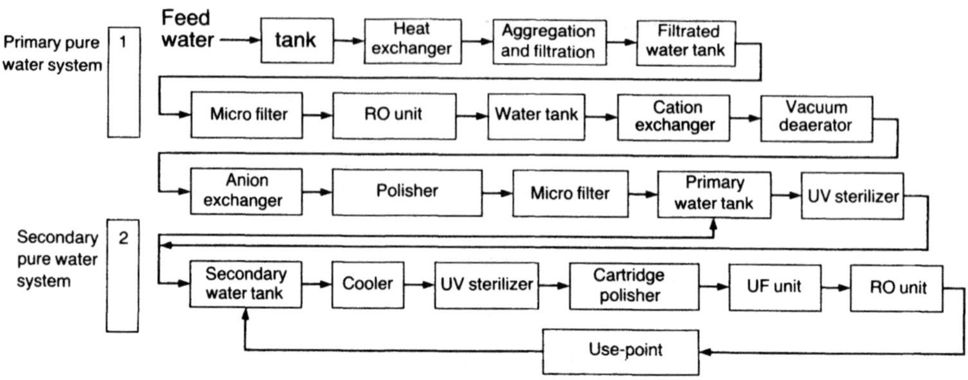

Figure 7.5 A process for production of ultra-pure water [76].

and absorbed into ultra-pure water. Similarly. HCl, HF and ammonia used in semiconductor processing are made by absorption of metal-free gases into water. The process flow sheet consists of micro-filtration, distillation, absorption and storage. The piping and equipment are made of glass or glass clad metal.

Secondary contamination from workers during filling, storage and transportation is a significant problem. It is avoided by automating clean room operations. Other high purity reagents are also supplied using fully automated systems.

Oil driven vacuum pumps are used for the production lines in processes such as CVD, etching and washing. The exhaust gas contains a large volume of the reactants in addition to the products (gases, liquids and solid particles) because of the low conversion. The waste gases contain oxides, halogens, hydrides and metal oxides, which cause large changes in the viscosity of the oil because of polymerization or degradation. Vacuum oils made from dialkyldiphenylether, alkynaphthalene and fluorinated hydrocarbons are used in vacuum pumps.

There are of the order of ten chemical companies producing reagents in Japan, with a market value of 80 billion yen per year.

7.2.4 Ultra-pure water

In all VLSI manufacturing processes washing with ultra-pure water plays an important role. The total amount of water needed to produce a 6 inch (15 cm) wafer is about 1.2 m^3. The specification which must be currently met for ultra-pure water is 18 MΩ specific resistance, less than 20 particles per ml, less than 0.01 bacterium particles per ml and 0.1 mg/l of total organic carbon (TOC) with a requirement to reduce this further.

A typical flow sheet of a process is given in Fig. 7.5 [76]. The feed water is treated with aggregated solids, followed by micro filtration, reverse osmosis (RO), degassing, ion exchange and ultra filtration (UF). The primary ultra-pure water is fed to a secondary production line and purified further at the point of use in a UV-sterilization unit and polishing with ion exchange, UF and RO. To prevent contamination, a part of the secondary ultra-pure water is recycled continuously through the pipework. About 70% of the secondary stream is recycled and the rest is discharged as effluent. Environmental pressures are giving rise to calls for a completely closed water system.

To try to reduce costs further and improve the quality of the ultra-pure water, further efforts are being made to develop a low pressure (*c.* 15 bar) RO system, using new UF membranes which have high heat resistance for thermal sterilization and contaminant-free piping materials. The market for ultra-pure water in Japan is about 35 billion yen (in 1985) and the market is dominated by three major companies which together supply 95% of the demand.

7.3 ENVIRONMENTAL CONSIDERATIONS

There are two environmental aspects of concern in VLSI operations. The first is to ensure that the atmosphere during processing of the chips is clean. The second falls into what is usually referred to as the domain of environmental pollution control which is concerned with problems of gaseous and liquid discharges and preservation of the local external environment.

7.3.1 Minimization of pattern defects

A major consideration is concerned with yield so very high priority is given to finding a means of increasing it while maintaining a high quality product at low cost. The operations which give rise to rejects are pattern defect 70%, processing, 15%, metal pollution, 2%, planning 3%, and others, 10%. The most important factor which needs to be considered to improve the yield is reduction of pattern defects. These are caused by particles sticking on the wafer during processing. Figure 7.6 illustrates how rejects are caused by particles sticking on the surface of the wafer. The extent of the defects on the wafer depends on the way in which the particles impact on the surface as well as on the composition and diameter of the particles. For example, a salt particle on a wafer causes a defect which is more than four times the particle diameter. Particles which originate from the heat treatment processes involving diffusion and oxidation enlarge the initial diameter considerably. Table 7.3 gives a classification of the effects of the particles in terms of the causes. Generally speaking, the diameter of the particles which would give rise to

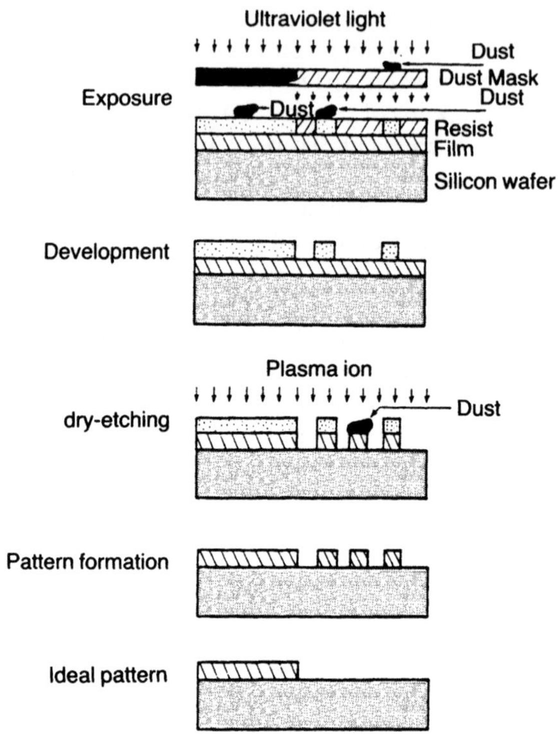

Figure 7.6 Models of pattern defect occurring by dust.

Table 7.3 Influence of sticking particles

Polishing (silicon, glass)	Scratch
Mask manufacturing	Coming out, remaining
Resist	Radiated floating of resist film
Ion pouring	Non-uniform silicon impurity density
Film formation (epitaxial or chemical vapour deposition)	Coming out, abnormal growth, floating
Oxidation, diffusion	Carbonization, sintering → baking

unsatisfactory devices is of the order of one tenth the width of the narrowest pattern line. Figure 7.7 illustrates the stages and indicates the acceptable range of the particle diameters and detection limits for the monitoring instruments for a 1Mbit wafer. Particles of more than 0.1 μm diameter must be removed and those down to 0.05 μm should be removed for 4Mbit. Further improvement in detection and removal of particles is still required.

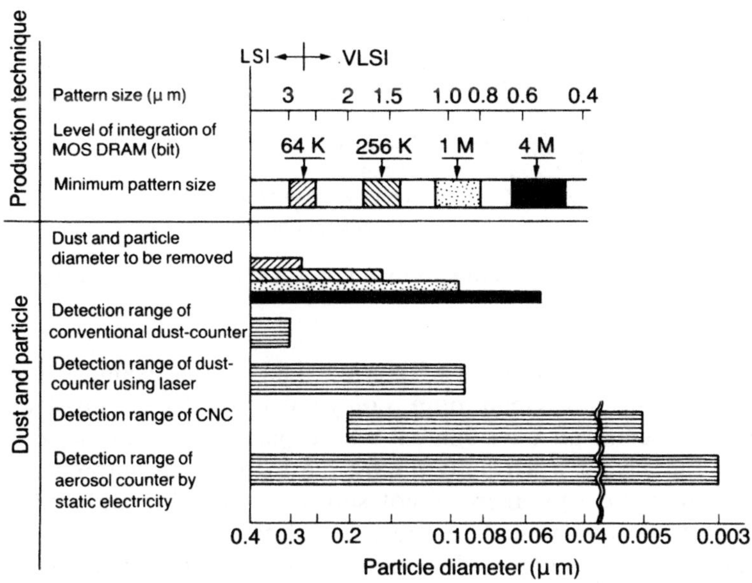

Figure 7.7 Development of LSI and ranges of particle diameter to be removed.

Table 7.4 Sources of particulate generation and amounts of particles in clean room

Particle cenerator	*Counts of particles (>0.05 m) (No./ft³)*
Below HEPA filter	1–10
Operator	10^3–10^6
Inside of processing equipment	10 –10^6
Process handling	10 –10^3
Clarifing water	10^5–10^6
Medical fluid	10^5–10^8
High purity gas	1–10^5

During VLSI manufacture, everything touching the wafers, directly or indirectly, must be clean. Table 7.4 shows the sources of particles and the concentration level typically found in a clean room. These can be classified as: operators, the inside of the equipment and utilities. High quality products require a clean air environment, utilities for wafer treatment equipment (high purity water, high purity gas), resist developers, special gases, skilled operators and automated equipment.

Figure 7.8 gives an example of the factors which can be expected to

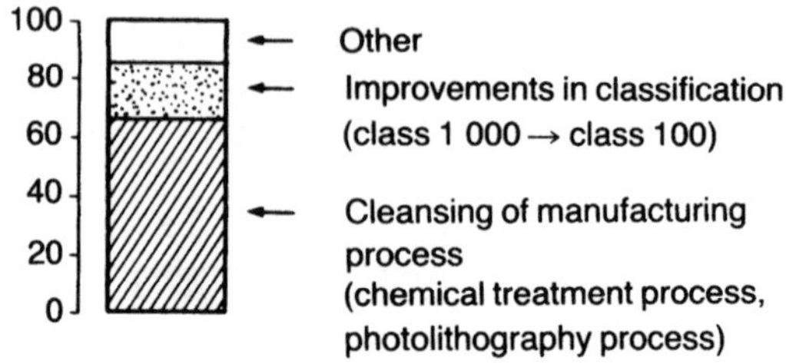

Figure 7.8 Contributing factors to the improvement of manufacturing yield.

lead to an improvement in manufacturing yield. In this example, it can be seen that the air in the clean room is only one of the factors involved. It is important to creat an environment where particles do not stick on the wafer. Generally, it is virtually impossible to build an automated clean room because the equipment is not sufficiently reliable and will in any case need to be maintained. A practical solution is to divide the system into the wafer processing inside equipment and the work involving operators, including maintenance [78]. The problem then is how to remove the wafer automatically from the airtight box into the carrier and finally into the equipment [79].

Steps are now being taken to develop machinery which will handle wafers cleanly. For example, belt conveyers which have been commonly used for moving single wafers are now being replaced by pick and place systems which avoid picking up particles by rubbing of the wafer on the belt [80]. Other methods being introduced which reduce the risk of particles being formed are based on linear motors and magnetic levitation.

However, there are still unexplained phenomena concerning the mechanism of particle contamination of wafers from the residue derived from physical or chemical processing. Secondary reaction products from the oxide furnace and dry etching stick on the wall of the equipment and are then released. Also particles from inside the equipment are generated by rubbing of components. As the particles approach the wafers, there is the possibility of impact and sticking. Clearly, an important consideration is that equipment should generate as few particles as possible.

7.3.2 Clean room conditions

The clean room should be maintained at 23 ± 1°C, with close control of humidity. More than half the electrical power for the whole of the manufacturing process is used in providing air conditioning for the clean room. The greater the degree of filtration, the greater the need to reduce energy consumption. A clean room works for 24 hours and so has a cooling load throughout the year. The load per unit area is more than twice that needed for office buildings (in case of 256 Kbit DRAMs). Of this 45% orginates from heat generation from the machines. To reduce the cooling load, the equipment should be cooled using outside air. Other energy savings can be made by minimizing re-circulation and controlling the airflow [81].

7.3.3 Liquid effluents and waste gases

The amount of chemicals used ranges from 10 t to 1 g per day. These materials, which must meet high purity standards, require special handling. For example, some materials need to be withdrawn, purified and re-circulated, as is the case for very high purity water or photo-resists. Some waste is sold to other industries and so does not require purification. Other discharges arise from processing equipment in the factory.

Table 7.5 shows the acid, alkaline and organic compounds which are discharged during wafer processing. Generally, such materials should be treated in the factory because they cause external environmental pollution problems. The methods used are those found in the chemical industry and therefore pose no problem. For gaseous discharges, two methods are used namely absorption and adsorption on to solids (mainly activated charcoal). Liquid effluents use neutralization, gas-liquid contacting and adsorption.

Recently, attention has been drawn to the problems of pollution of underground water by organic chlorine compounds and it is clearly a major social problem. According to data from 303 cities, towns and villages in four urban and rural prefectures in Japan in 1986, 5.2% the percentage of the wells exceeded the international recommended level of trichloroethylene (less than 0.03 mg/l) and 3.9% for tetrachloroethylene (less than 0.01 mg/l) [82, 83]. This is a sharp increase compared with data for 1984 and 1985, although a strict comparison is difficult because the tests were conducted in different places.

In looking at how environmental pollution issues have grown, the starting point can be taken as the leakage of trichloroethylene, trichloroethane, tetrachloroethylene, and toluene into the underground water from the semiconductor factories of sites in San Jose in California, in 1980. This happened because of cracks in nine underground storage tanks. In Japan, four factories exceeded the limits for trichloroethylene, according to a study of drains by the Environment Agency in 1984. This material is used in cleaning for semiconductors and as a degreaser for metals, as well as in dry cleaning. In 1986, production in Japan was 71 332 t, with imports of 1775 t and exports of 8833 t. The requirement for semiconductor processing was about 3000 t in that year. For comparison, the demand for tetrachloroethylene was 91 661 t, of which 17% was used as a cleanser and a degreaser and 30% for dry cleaning.

Table 7.5 Wastes discharged during wafer processing

Type of waste	*Main components of waste*		*Main generating process*
Waste acid	Waste fluoric acid (HF, NH_4F) other waste acid (sulphuric acid, hydrochloric acid, nitric acid, etc.)		Etching, clarification ditto ditto
	Waste acid containing acetic acid		ditto
Waste alkali	Organic alkali		Lithography
Waste oil	Waste of organic solvent	Xylene, toluene, butylacetate, etc.	Lithography
		Acetone, isopropylalcohol, trichloroethylene, tetrachloroethylene, methanol, etc.	Clarification Assembly
	Vacuum pump oil	Arsenic, phosphorus, oil containing boride	PVD, CVD Dry-etching Ion pouring
Arsenic or chromic waste liquid	Waste liquid of etching clarification (arsenic, hexavalent chromium)		Etching Ion pouring Evaluation etching

While there is some question as to how much the picture is clear with respect to VLSI manufacturing, waste is produced in metal processing and dry cleaning. Investigations by cities, prefectures and public agencies aimed at providing details of the problem has suggested many explanations as to the causes of pollution. The Environment Agency started an investigation into the contamination of underground water in 1984. This is an environmental issue because it is likely that the organic solvent may have diffused into the air and then be subsequently deposited on the ground, leading to contamination of underground water.

Since October 1986, the Environment Agency in combination with the Ministry of International Trade and Industry, the Ministry of Health and Welfare, and the Ministry of Labour has conducted a survey of drains, waste gas, solid waste product processing and working environments. It has also investigated the use of chemical elements in VLSI factories in all parts of Japan and looked at techniques for eliminating waste matter. In February 1989, the Central Council on the Environment in Japan published a draft report entitled 'The ideal way of preventing contaminants in underground water'. The expectation is that this will provide a basis for formulating a policy regarding industrial accidents and pollution in the semiconductor industry.

8

Process engineering aspects of VLSI production

In the preceding chapters, the history, unit processes and scope of VLSI technology have been described. Chapter 5 was devoted to a case study in order to bring out the details of the process technology. This chapter will focus on how chemical engineering can be applied to VLSI technology.

Adopting the same process as the case study, it can be seen that the wafer treatment process involves oxidation, ion implantation, washing, etc. which is similar to operations encountered in chemical processes based on reaction, unit operations, distillation, extraction, drying and so on. As a first step in analysing how chemical engineering principles can be applied, the representative unit operations need to be identified.

Table 8.1 summarizes the problems to be solved in terms of chemical engineering topics encountered in the manufacturing process, based on information from experts, and draws attention to the fundamental problems to be addressed. These can be matched against the relevant areas encountered in chemical engineering. Since the scale of integration in VLSI changes significantly every 2 to 3 years, the outline given here should be seen as being appropriate to 1–4 megabyte devices based on a minimum lithographic line width of less than 1 μm.

Generally speaking, the issues to be resolved depend on establishing a compromise between fineness of scale, accuracy and productivity. It might be thought that product quality is the most important consideration, with production rate being regarded as less important, because the decisive factor is device performance. In fact, the requirement is for a process which satisfies both quality and rate simultaneously. The following sections examine the individual issues associated with each of the unit operations.

8.1 BASIC ISSUES IN PROCESSING

8.1.1 Crystallization

There is an ever-increasing demand for larger wafers in order to reach higher productivity: micromachining for higher integration will also be required in the future. The following considerations are relevant:

- production of very pure high quality crystals with no defects;
- good control so as to ensure uniform doping;
- development of technology and equipment for production of large diameter single crystals; and
- processes for wafer shaping including flattening and gettering.

As already seen in section 6.1 the technology to produce dislocation-free single crystals has been directed specifically at VLSI processing. However, micro-defects or impurities are now becoming a problem. The formation of micro-defects in the silicon wafers during thermal treatments is closely related to the existence of carbon and/or oxygen atoms. The carbon atoms originate in the production of metallic silicon while oxygen arises from the silica crucible as a result of gradual dissolution in the silicon melt.

Purification of silicon to remove the carbon atoms and the mechanism of inclusion of oxygen atoms dissolved from the crucible need to be considered in detail. Convective flow in the melt is reduced by application of magnetic fields so that transfer of oxygen from the bulk melt to the growing crystal surface is prevented. Optimum conditions can be determined using simulation which needs to take account of changes in the physical properties due to the magnetic field.

Growth of uniform single crystals is crucially dependent on control of the conditions. The pull rate cannot be analysed in terms of heat transfer alone. Consideration of the growth mechanism and the effect of impurities and dopants is essential to ensure high quality single crystals. Unsteady state simulation can be used to determine the preferred growth conditions which must take account of both heat and mass transfer, surface interactions and fluid flow. A particularly important factor to be considered is the variations in the melt conditions near the growing interface.

For growing large diameter crystals, the control of convective flow and heat transfer in the melt induced by the rotation of both the crystal and the crucible is a particularly difficult problem compared with that of small crystals. New methods need to be developed to control the diameter, provide precise temperature measurement and detect the position of the interface. Simulation techniques are very effective for attacking these problems.

Both the micro- and macro-scale flatness of the wafer are essential in

Table 8.1 Subjects in Chemical Engineering

	Crystallization technology	*Oxidation technology 5.2.2.*	*Measurement of film thickness in ion implantation and annealing technology 5.2.3*	*Formation of films (CVD, sputtering) 5.2.4*
Topics	1) Defect-free crystals 2) Uniform distribution of impurities 3) Control of crystal diameter 4) Flattening of wafer 5) Gettering	1) Formation of thin films 2) Reaction at low temperatures 3) High production rates	1) Shallow implantation at high ion concentration 2) Defects in crystals and recrystallization 3) Short-time annealing	1) Planarization 2) Trench filling stuffing 3) Selective growth 4) Development of thin films having high dielectric constant 5) Epitaxial growth
Reaction engineering	1,5) Reproduction of carbon and heavy metals 5) Optimal concentration of oxygen	Optimization of oxidation process 2) Mechanism of defect formation	1) Generation of high ion concentrations 2) Mechanism of generation of crystal defects and recrystallization	1) Physical properties of film 2) Mechanism of film formation 3) Optimal reaction conditions 4) Reaction system and reactor 5) Development of plasma for low temperature operation
Heat transfer	1,2,3,5) Heat transfer in liquid or solid phase 1,2,3,5) Natural convection 4) Maintaining uniform temperature of wafer	Uniform temperature profiles	1,3) Prevention of temperature increase in wafers 1,3) Maintaining uniform temperature profile	Control of temperature profile

Fluid dynamics	1,2,3) Flow of high viscosity fluids 1,3) Fluid dynamics in a magnetic field	Fluid dynamics of reactant gas	Solid phase diffusion 1,2) Distribution of impurity species	Fluid circulation patterns and film growth rate Distribution of gas supply
Mass transfer	5) Diffusion of impurities in the solid phase 2,3,5) Concentration of impurities	1) Distribution of impurities on $Si\text{-}SiO_2$ surface		Diffusion in gas phase and/or on the surface
Separations technology	1,2) Removal of small amounts of impurities 1,2) Purification of raw materials			Purification of feedstock
Powder technology				
Process systems engineering	1,3) Control of crystal growth rate 1,3) Control of wafer diameter	1,2) Control of temperature distribution	3) Temperature control	Control of temperature profile
Simulation	Estimation of shape during pulling or crystal fluid dynamics and heat transfer in CZ method	Calculation of temperature distribution and impurity concentrations Estimation of film thickness	Calculation of temperature distribution and impurity concentrations	Film thickness Distributions of composition and internal stress
Other	Measurements of concentration and high temperatures	Film thickness measurement		Physical properties of film Materials selection

Table 8.1 *(Continued)*

	Lithography 5.2.5	*Etching* 5.2.6	*Cleaning (including drying) 5.2.1*	*Process design and management*
Topics	1) Resolution 2) Alignment accuracy 3) Accuracy of depth of focus	1) Preventing destruction of gate 2) Improvement of aspect ratios of resist and mask 3) Accuracy of etching 4) Evaluation of trench surface 5) High-speed etching or trench	1) Effective cleaning and drying methods 2) Measurement of very small dust particles 3) Purity of chemicals, gas and water 4) Design of clean room 5) Development of effective vacuum system	1) Development of management system 2) Development of factory automation system 3) Instrumentation and evaluation 4) Optimal equipment layout
Reaction engineering	3) Analysis of interaction between exposure, sensitization and development	1,2,5) Mechanism of anistropic etching 1,2) Selection of reactive gas 2,3,5) Shape control in etching	1) Reaction mechanism	
Heat transfer engineering	Maintenance of uniform temperature	2) Temperature profile	1) Prevention of spots on surface during drying	
Fluid dynamics	Uniform coating of resist	2) Gas circulation	1,4) Circulation of liquid and gas	
Mass transfer		1,2) Gas diffusion	1) Mechanism of cleaning 1) Prevention of spots on a surface during drying	

Separations technology			3) Removal of small impurities 1) Development of cleaning and water treatment methods	
Powder technology			Dust generation mechanism 4) Development of removal methods 2,4) Measurement of very small dust particles	
Process systems engineering		Estimation of end point during etching	4) Optimal design	1) Production control 1,2) Process system design with redundancy
Simulation	Estimation of cross-section of resist	Estimation of end point during etching Estimation of cross-section of trench	Circulation of liquid and gas motion of dust	Operational status of process
Other	1,2) Interaction between etching and ion implantation and thickness and solidity of films		5) Dust protection during pump down	

order to ensure precise micro-machining for transferring the mask pattern to the wafer. Not only is it necessary to improve the lapping process during wafer formation and the measuring technique for flatness, but also the bending of the wafer caused by thermal expansion during the LSI processing must be overcome. Bending is a particularly difficult problem since there is evidence that local oxygen segregation is partially responsible. With large diameter crystals, there is a temperature distribution in the wafer during fabrication and in the subsequent treatment process. This calls for careful temperature control during temperature treatment of the wafer.

8.1.2 Oxidation

There are three problems associated with the oxidation technique, namely:

1. the thin film, i.e. less than 10 nm (high voltage resistance, high quality, trench structure);
2. low temperature (increasing voltage resistance, uniformity of temperature); and
3. high speed.

There has been considerable research into how to produce thin films to form a capacitor dielectric which will retain electric charge for use in memory. The thicknesses used are now down to 10 nm. This film must be able to resist electrical breakdown of the dielectric caused by the applied electric field which can be as high as 5 MV/cm in the case of 5 V supply. A design figure for the thickness of the capacitor dielectric film is typically about 8 MV/cm for a 4–6 nm film. The gate dielectric film thickness is usually about 15–20 nm. It is important to be able to produce capacitor films of uniform thickness and quality, particularly with respect to ripples in the surface and the shape of the trench. This must be done for discs up to 8 inches (20 cm) which requires a redistribution of the impurities and stacking faults caused by the oxidation process, while at the same time reducing the warp and stress in the wafer.

To achieve the necessary immunity to voltage damage and form a high quality planar film or a three-dimensional structure, the oxidation temperature should be as low as possible, preferably 800 ± 1°C. However, at low temperatures the rate of oxidation is slow so a compromise is necessary. Research is being conducted into the use of high pressure and plasma oxidation, both of which are showing promise of leading to improved rates.

High pressure oxidation is now desirable because a technique for forming a thin oxide film has been developed [84–6]. The next step requires the thin oxide film to be well-formed which is dependent on good, safe operational procedures which ensure uniform temperature and density during processing. This presents a considerable challenge in equipment design. It requires a well-defined relationship between cause, in terms of prescribed operating conditions, and effect, as expressed by the resulting film characteristics. The design of the equipment has to balance reaction and transport processes for heat and mass with fluid flow at the interface of the gas and liquid substrate, as well as taking some account of the solid properties at the micro- and molecular levels.

Plasma oxidation is still at the research stage and requires further investigation of the oxidation kinetics, particularly in complicated structures, so as to ensure a uniform oxide build up on the disc.

8.1.3 Impurity implantation (doping) and annealing

Reduction of line width to the critical dimension calls for the following:

- shallow implantation of ions at high concentration;
- reduction in crystal defects and better re-crystallization; and
- improvements in short term annealing procedures.

With a narrower line width, the region of high impurity concentration will be thinner than before, so the need for high concentration ion implantation with tailing control over the doping region becomes important. There are two approaches to this problem:

1. development of ion implantation using low energy beams at high current density; and
2. implantation of impurities from the Group III elements instead of boron.

In the case of (1), the ions are generated in a chamber and accelerated by a low energy field to form a thin impurity layer in the wafer. To satisfy (2), it is necessary to reduce the acceleration energy. To implant molecular ions in a wafer, the impurities must have relatively small energy. Also, some ions are decomposed before they are implanted and this makes it difficult to control the composition precisely. This is a typical reaction engineering problem concerned with the generation of stable ions at high concentrations.

High concentration ion implantation causes crystal lattice defects which annealing can only partially repair. The relationship between the number of crystal defects in dielectric films and their electrical properties is not fully understood, that is, it is not clear whether defects cause a deterioration in electrical properties. It is therefore necessary to establish whether there is an upper limit to the number of defects that can be present before deterioration occurs. Another major problem is that annealing causes the implanted active impurities to form clusters, and this results in their uneven distribution throughout the film.

As the line widths become narrower, layers previously formed on the wafer can be affected by long term annealing at high temperature. The development of short time laser annealing has overcome this problem but while this produces a single crystal layer in a short time the defects still remain. In addition, it is important to be able to predict how far the impurities have diffused into the solid so as to enable optimal conditions to be determined. There is also a need to be able to estimate what effect the operating conditions have on the parts already formed to enable the quality of the device to be controlled.

Other matters which need to be considered are:

- the difficulty, arising from presently available equipment, of ensuring there is a uniform distribution of impurities as the diameter of wafers increase, and
- the removal of heat generated by collision of ions with the surface of the wafer.

8.1.4 Thin film preparation

The issues arising here are

flattening the surface,
filling the trench,
selective deposition,
preparation of films with a high dielectric constant, and
epitaxial growth.

The ever increasing growth in scale of integration causes the surface of the wafer being processed to become more and more uneven. The first three items above all give rise to surface unevenness because of the high aspect ratio.

Two different approaches are possible in flattening the surface. As shown in Fig. 8.1(d) and (e), one technique is to try to deposit a flat film over the uneven surface: another one is flattening of the film after growth. The former includes bias-sputtering, which is an improvement on CVD and the spin-on-glass method. Bias-sputtering involves biasing the substrate negatively. The high energy ions bombard the uneven surface and cut off the side wall of the trench which makes it possible to penetrate deeper.

Improved CVD processes have been developed successfully by using alternative gaseous source molecules such as tetraethoxy silane (TEOS). Pyrolysis of TEOS followed by oxidation with ozone results in good quality step coverage. It often results in conformal growth, as shown in Fig. 8.1(c). Filling up of a trench, as shown in Fig. 8.1(d) also occurs. An analysis of the mechanism which determines step coverage quality and the

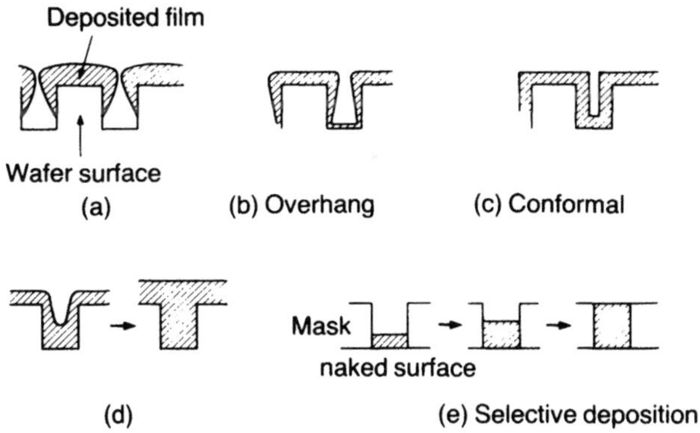

Figure 8.1 Categorization of deposition profiles on a surface of high aspect ratio.

search for an optimum process can be done using standard chemical engineering techniques and will be discussed in 8.2.2. Spin-on-glass produces an SiO_2 film formed by the polymerization of a spin-coated layer containing monomer molecules. There is a general trend in VLSI towards dry processes rather than wet, but the step coverage quality with wet processing is superior to that in dry processing and this has opened new applications for the spin-on-glass method.

In parallel with the efforts to form uniform films, a deposition process followed by etching, called 'etch-back', is being developed. The process is outlined in Fig. 8.2. Although the step coverage quality is improved, this is at the cost of having to repeat several steps in the processes. Unfortunately, VLSI technology is beset with such problems but the success of the techniques is crucial in establishing a viable process.

The filling of a relatively deep trench of several microns to form a trench type capacitor is a good example of the problems which must be overcome. The resulting capacitor should have an area large enough to accumulate an electric charge greater than that generated by radiation from radio-active sources in the materials. This is a well-known problem in chemical engineering since it is analogous to the concept of catalyst effectiveness [88] or absorption enhancement factor [89], where reactant molecules diffuse into a porous catalyst or a liquid while at the same time reacting. The progressive depletion in reactant effectively reduces the effectiveness of the reacting volume. The analysis of the capacitor problem can be analysed in these terms as will be shown in 8.2.1.

A new approach to formation of thin layers is being developed which promotes selective deposition on a substrate. The difference in the chemical nature of those parts of the substrate which are exposed and covered (masked) causes selective growth. Selective and epitaxial growth

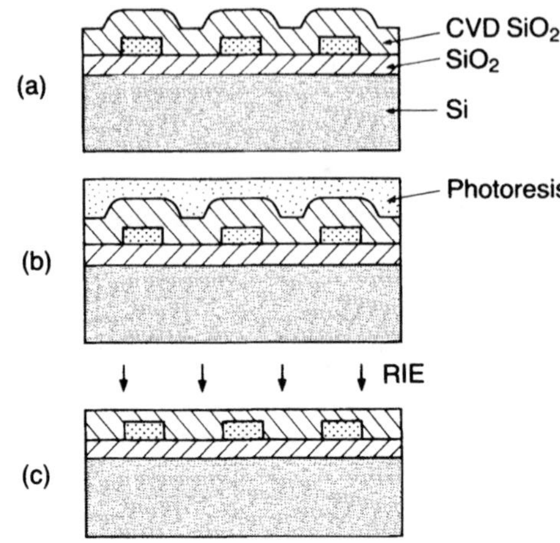

Figure 8.2 Schematic of etch-back method.

of silicon on a wafer masked partially by SiO_2 was first reported in 1962 [90]. Tungsten, aluminium as well as gallium arsenide have also been shown to deposit selectively [91]. Tungsten is regarded as a possible replacement of aluminium for metallization. When the surface of the silicon is partially masked by SiO_2, gaseous WF_6 decomposes to form W on the silicon surface. The ability of silicon to reduce WF_6, while silicon dioxide cannot, is believed to be responsible for the selective nucleation on these materials.

The nucleation reaction on silicon can be written as

$$Si + WF_6 \rightarrow SiF_4 \text{ (desorption)} + WF_x \text{ (nucleation)} \qquad (8.1)$$

The growth due to reaction of W is

$$WF_6 + 3H_2 \rightarrow W + 6HF \qquad (8.2)$$

This selectivity makes is possible to deposit a film of W in a pattern such as that shown in Fig. 8.1(e). It is based on the fact that the probability that nucleation occurs on SiO_2 is low. Even if this is as low as one in a million it will have an effect on the yield of 1 megabyte chips. Indeed, it is this problem, as well as the possible damage of the silicon surface due to reaction (8.1), which has prevented selective deposition being used in practice.

Selective deposition does suggest an alternative approach, though there are still many problems remaining to be overcome. Rather than following the usual practice of forming a film over the entire surface, followed by lithography to etch the surface, a film is formed only on the desired parts.

A solution to this problem would represent a breakthrough and there are encouraging indications that it will be achieved.

Capacitance depends on the product of the area of the electrode and the dielectric constant of the materials and the spacing between electrodes, so if the area and spacing are fixed by the dimensions, the determining factor is the dielectric constant of the spacing material. This should be as high as possible and research is underway to develop films of Ta_2O_5 and other oxide materials with a thickness of the order of 10 nm for this duty. It is essential that the films are defect-free and highly resistive to electrical break down, even under a very high field strength.

There have been frequent reports of revolutionary developments which always gives rise to optimism. Unfortunately, there is a large gap between success in the laboratory and its translation to industrial production. VLSI processing provides a good example where success can be compromised by low efficiency in any of the memory cells.

There is a need to develop a low cost process for growing epitaxial silicon since a superior surface epitaxial layer is crucial in increasing the level of integration. It also plays a part in forming p-n junctions which isolate individual cells. This requires the problem of high temperature autodoping during epitaxial processing to be solved. Any diffusion of the dopant from the substrate to the p-n interface causes inferior performance so a low temperature process would help to overcome this.

In conclusion, thin film deposition processes contain many problems yet to be solved. Chemical engineering techniques can provide a powerful range of tools which make use of knowledge of diffusional and kinetic phenomenon to assist in this task.

8.1.5 Lithography

The design rule (section 4.1.4) for a line width of 1.3 micron implies a maximum memory size of 1 Mbit DRAM (this can be increased to 4 M if the width is reduced to 0.8 micron). All functional elements and circuit connections must therefore be less than 1 micron which is basically determined by the accuracy of the lithographic process. The problems in improving this precision can be listed as follows:

1. increase in resolution (preparation of high precision defect-free masks using shorter wave length light sources and improved photo-resists);
2. more accurate aligning; and
3. complete removal of resists.

Accurately defined line widths are essential in achieving higher density integration. Currently, UV light is used but there is a limit in the minimum possible width of 0.8 micron, and hence on the maximum memory size of the 4Mbit DRAM. This limit has been reduced to 0.5 microns by using

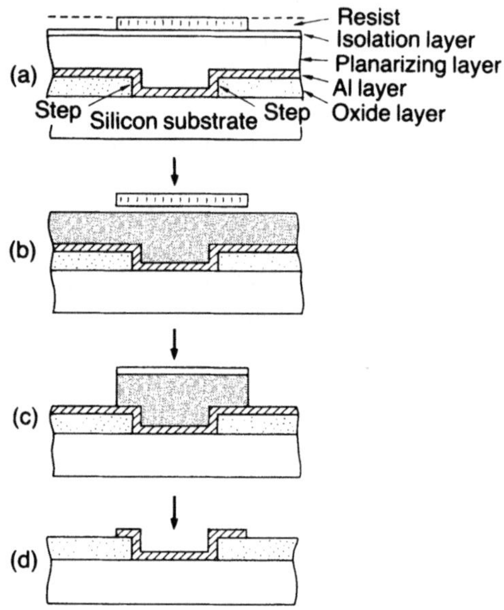

Figure 8.3 Multilevel etching: (a) projection and resist-patterning; (b) etching of isolation layer; (c) etching of planarizing layer; (d) etching of Al layer and removal of planarizing layer.

stepper technology. Future improvements will require changing to shorter wave lengths, such as electron beam (EB) or X-rays. To achieve better resolution, however, it is first necessary to have more precise masks and to overcome the technical problems of flattening wafers and the photo-resists. From a chemical engineering point of view, the problems to be solved relate to finding photo-resists which have a higher selectivity and to understanding the mechanisms which determine the interaction between exposure and developing time. There would also be benefits from having improved contact between the resist and wafer.

Efforts are now being made to extend photo-techniques to make use of multi-level resists. An example of this would be in the etching of an Al layer on a stepped silicon dioxide surface, as shown in Fig. 8.3. The planarizing lower resist layer is of sufficient thickness to cover the steps on the outer surface. The middle and top layers are then spread over this. Using lithography for the top resist layer (a), the pattern of the middle layer becomes a mask for the lower (b). Subsequent etching of the planarizing layer leaves the desired pattern on the Al layer (c). Final etching of the Al layer and the removal of the resist yield an accurately micro-machined element (d). To do this requires a highly selective photo-resist. A quantitative analysis depends on knowledge of the mechanism and rate of etching and this will be discussed later. The crucial factors which need to be considered in order to achieve higher precision during processing and increasing the throughput are increased writing area, shorter wave length light sources, more precise alignment of the resist, ability to handle large wafers, high etching speeds, automatic exchange of masks (reticules) and elimination of dust.

8.1.6 The etching process

Dry etching is preferable in VLSI processing because it has advantages over the wet process in not having to treat the waste liquid. It also makes it possible to create anisotropic structures, which are required for fabrication micro-machinery. Further development depends on solving the following problems:

- prevention of gate breakdown;
- improvement in the selectivity of resists and mask material;
- improved precision; and
- evaluation of the state of cleanliness of trench surfaces.

RIE is the most commonly used dry etching process but gives rise to problems cause by ion impact which results in breakdown of the gates and causes lattice defects. To overcome these, a light-excited etching technique has been suggested. However, the generally favoured approach is to try to improve current technology by using lower energy beams.

In order to increase the selectivity ratio of the mask material it is necessary to increase the accuracy of the etching. This means improving the stability of the mask materials with respect to the reactive gases and the selective etching ratio of the mask in multi-layered resist methods, including etching of SiO_2 layers on silicon wafers. To do this, the etching mechanism for different combinations of mask materials and reactive gases should be understood and the kinetics of the process controlled. In particular, to prevent over-processing the detection of the end point for etching needs to be identified. A possible approach is to use a spectroscopic method. It is also necessary to assess the shapes after etching as well as the cleanliness of the surface of the trenches.

Etching dusts (foreign particles) may cause serious problems since they behave as masks and give rise to anisotropic etching. The dust arises from precipitates formed by plasma polymerization of some of the reactant gases or from the reaction products. Sometimes, however, the dust comes from other parts of the equipment. To attempt to optimize the operating conditions and improve the performance by better mechanical design, a knowledge of reaction mechanisms is needed to understand how the problems arise from the dust.

8.1.7 Cleaning (including rinsing and drying)

The following list highlights the most pressing problems:

1. development of effective rinsing and drying methods for the scale of operation;
2. finding a means of determining the effectiveness of rinsing and drying procedures and measurement methods to detect foreign molecules;
3. availability of higher quality chemicals, gases and water;
4. optimum design methods for clean rooms;
5. effective methods for producing high vacuum;
6. reduced contamination of equipment;
7. overcoming problems arising from CFCs (chlorofluorocarbons).

The production of 4 Mbit DRAM memory calls for good rinsing and drying procedures for trench construction to ensure high quality films are formed. The liquid residence time should be sufficient to prevent re-adhesion of contaminants in the trench and formation of blotches because of uneven drying. An important part of this is having a clear picture of the micro- and macro-reaction mechanisms of the contaminants and chemicals in terms of the surface chemistry. There is an urgent need for an effective dry cleaning method to be developed.

Measurement of impurities

Non-destructive and non-contacting methods for determining whether a surface is clean have to be based on analytical instruments which measure and identify components rapidly.

Quality of materials

The surface of the wafer is very active and any impurities in the water, chemicals or gases which come into contact with it have a high probability of sticking. Consequently, it is essential to have the extremely pure materials so that a high quality film is formed. This applies particularly to water, which must also be cheap, since about 1.2 m^3 per wafer is required.

Clean room

High efficiency dust collectors are essential. This requires an understanding of the movement of submicron dusts in the clean room. To achieve high removal efficiency for the clean room, the location of exhaust fans needs to be optimized.

Generating high vacuum

During film formation and ion implantation, various levels of vacuum are required. The exhaust gas contains unreacted molecules and solid particles of Si, so it is important to develop a high efficiency evacuation system.

Reduction in contamination

To reduce dust formation in the equipment and prevent adhesion of dust particles, including those due to re-scattering from the walls, it is necessary to understand how the particles migrate and the mechanism of adhesion on the wafer surface. Without this it is impossible to improve the quality of the film and reduce the incidence of unsatisfactory lithography because of irregular etching.

Chlorofluorocarbons

CFCs have been identified as substances which reduce the level of ozone in the stratosphere. The use of R113 is being reduced and it will be withdrawn in the near future. Consequently, while they continue to be used, these chemicals should be recovered and recycled in the closed-loop system until alternative materials become available.

8.1.8 Design and management technology

The problems of design and management of the complete VLSI process are:

1. development of a process management policy;
2. development of an automated manufacturing procedure;
3. establishment of the instrumentation and evaluation techniques; and
4. optimal design of plant layout.

Management can be divided broadly into processing and production activities. Production management is concerned with planning, i.e. what products will be produced and when. The procedures should control not only the operation of the equipment but what products are being produced, including the processing priority of batches and the procedures to be followed during any break-down. Since these tend to occur frequently, it is crucial to be able to continue during maintenance. As noted in Chapter 5, several elementary process technologies are used as part of an iterative VLSI processing cycle. The individual operating conditions are generally different and it is possible to waste time tuning machines. Overall, the goal is to reduce the elapsed production time.

VLSI processing requires precise operation. To prevent mal-operation, and achieve good product quality while maintaining safe operation and a clean environment, it is important to anticipate the problems likely to be encountered by operators so they know the appropriate procedures to use. Such education programmes are routinely carried out so this can be a very effective way of introducing TQC (total quality control) procedures. In establishing operational strategies it is necessary to prepare detailed sequencing manuals, as is common for chemical plants. In addition the problems are very similar to those found in batch chemical processing, although more stages are needed in VLSI.

An important reason for introducing automation is to try to reduce the amount of dust since this can have a significant effect on yield. To be effective, automation should cover as many processing units as possible and even incorporate aspects of management in addition to instrumentation and evaluation, if maximum efficiency is to be achieved. Measurement and evaluation should be carried out after each processing step. The goal of any automation should be to increase the yield and reduce the elapsed time for processing.

Optimal design and plant layout are encountered in chemical processing since it is necessary to consider the supply of chemicals, gases, pure water and utilities, as well as the recovery and treatment systems for waste liquids and gases. In VLSI processing, almost all chemicals and gases are highly toxic, explosive and corrosive, although the quantities are small, except for those used for cleaning. The problem arises from the large number of waste products rather than the amounts.

An important design problem is how to determine the number of units necessary to carry out the processing and how many production lines will be required. This is typical of batch processing plants. The problem is to develop a systematic approach to obtaining a line balance. If this is done it is then possible to select the appropriate equipment and single (or batch) wafer processing stage as a sub-unit from which to assemble the complete process system as a superstructure. Since processing equipment often breaks down, some redundancy needs to be built into it so that production can continue if a fault develops. The difference as far as chemical plant is concerned is that the lifetime for equipment is only about 4 years, because of the rapidly changing technology and market. The continuous change in technology also calls for flexible design concepts.

8.2 OPPORTUNITIES FOR INNOVATION

8.2.1 Improvement in existing technology

Rationalization

VLSI processing is now highly developed in that fabrication of devices is controlled at a micrometre level in the semiconductors and even at the atomic level in terms of crystal defects. To understand the problems, it is necessary to examine the process in more detail. Even for the simplest stages, the relation between processing and the resulting wafer characteristics is often poorly understood. There is no quantitative understanding of the phenomena and so the processing units tend to be treated as black boxes. While there is some information as to how the product is formed under certain conditions. It is still not possible to explain why. In some measure this is because rapid changes are taking place in the technology and there is no time to undertake a detailed analysis of the process. This makes it difficult to formulate a strategy for the future because there is no fundamental understanding of what is happening. It is to be expected that super-VLSI technology will be even more complicated and need significant development effort.

There is considerable scope for applying chemical engineering principles in trying to go beyond black box models. Probing the fundamental mechanisms is necessary to provide the basis for the quantitative descriptions which will be needed to synthesize optimal equipment designs and identify the best operating conditions.

Many of the problem areas listed in Table 8.1 can be approached by simulation. For example, modelling of the large scale concentration and temperature profiles in the equipment in terms of the flow patterns, mass and heat transfers and chemical reactions is feasible. This is commonly used in chemical reaction engineering, although greater accuracy and precision are required in the case of VLSI. Thus, simulation of the processes occurring near or at the surface of a silicon wafer, as well as internally, requires the solution of the Navier–Stokes equation in conjunction with the diffusion and heat transfer. This provides information on how the reaction controls the quality of single crystals in terms of the defects in grain size of the polycrystals and the composition of the amorphous films. Because it is necessary to consider what is happening at the molecular level, the physical properties must take account of probability distribution functions for molecular interactions in addition to the classical continuum rate equations. This requires an extension of modelling procedures to accommodate the rather extreme situations where the descriptions must be expressed in terms of the micro- and macro-structures but must nevertheless be related to mass production of commodity materials.

To illustrate what is involved in simulation in the present context, we consider two examples: pulling single crystals of Si; and thin film preparation (trench filling).

Pulling single crystals of Si

The so-called Czochralski process is used for producing the majority of single crystals of silicon for VLSI use. The production of high purity polysilicon for crystal pulling is a standard chemical engineering operation (see section 6.1).

The process requires that the following important specifications are met:

1. no dislocations in the crystal: other defects such as impurities and holes must be avoided as far as possible;
2. the diameter of the crystal should be constant;
3. there should be a uniform prescribed concentration of dopants throughout the crystal; and
4. the amount of oxygen dissolved from the quartz crucible into the crystals should be minimized.

The objective is to find conditions which satisfy these requirements while meeting the desired product quality and device characteristics at high production rates and for large diameter crystals. The first step is to consider the crystal growth. During the CZ pulling stage the crystal growth is considered to be controlled by the thermal processes so the rate equations are governed by conduction in the crystal phase and conduction and convection in the melt. Heat transfer also takes place in the surrounding gas phase by radiation. Boundary conditions need to be specified so that the equations can be solved numerically using a suitable procedure. The computed solutions can be compared with plant measurements to determine whether the model needs to be refined or simplified. Figure 8.4 shows typical numerical solutions for flow patterns (left) and temperature profiles in the melt during crystal pulling. Figure 8.4(a) shows simulation of nonsteady flow pattern in a silicon melt in pulling a monocrystalline rod (CZ method) while Fig. 8.5(b) shows simulation of nonsteady temperature distribution in a silicon melt in pulling a monocrystalline rod (CZ method). In general, flow occurs in a melt of silicon in pulling a single crystalline rod. Because of considerably large values of Reynolds number, the regime of the flow is transient-to-turbulent in the case of CZ pulling of a large sized single crystalline rod of silicon. Nonsteady flow and temperature fluctuation occurring beneath liquid-to-solid interface affects the crystalline qualities such as segregation of dopants and oxygen and partial polycrystallization. Control to suppress these fluctuations is indispensable. The figures show the simulation results of pulling of a 6 inch rod, in which flow and temperature fluctuations are three-dimensionally calculated. High precision differential method is used to solve Navier–Stokes and energy conservation equations, self-consistently from the bulk melt to the boundary layer.

Such simulations are useful in understanding the phenomena involved. The model cannot expect to include all the factors involved. All that is necessary is that sufficient detail is included to ensure agreement with measured values in terms of properties such as growth rate, shape and

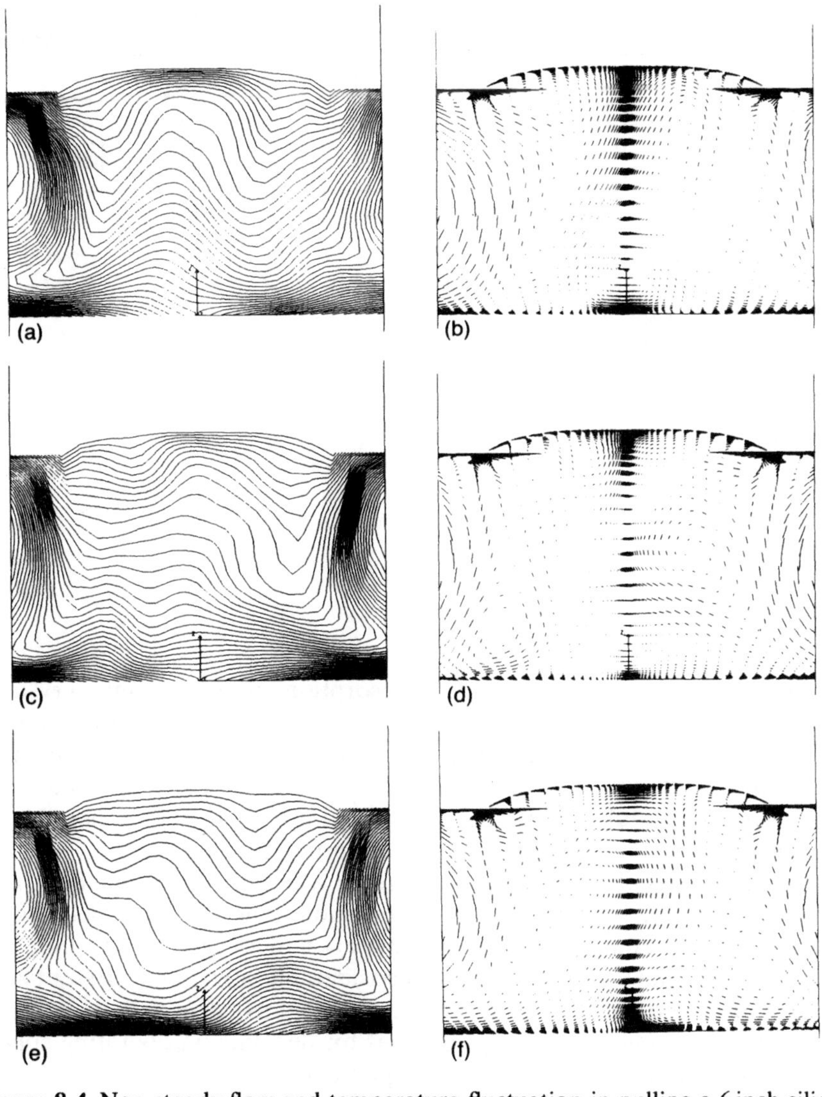

Figure 8.4 Non-steady flow and temperature fluctuation in pulling a 6 inch silicon rod [93] (a),(b) time = t_0 s; (c),(d) time = t_0 + 20 s; (e),(f) time = t_0 + 40 s.

pulling rates. When this is achieved, computational simulation can be carried out to explore a wider range of conditions than would be feasible on a plant or in the laboratory. This approach can also be used to explore process control problems by carrying out a sensitivity analysis for such effects as the gas phase temperature profile or the flow patterns near the interface on the crystal shape. It can also be used to identify the most

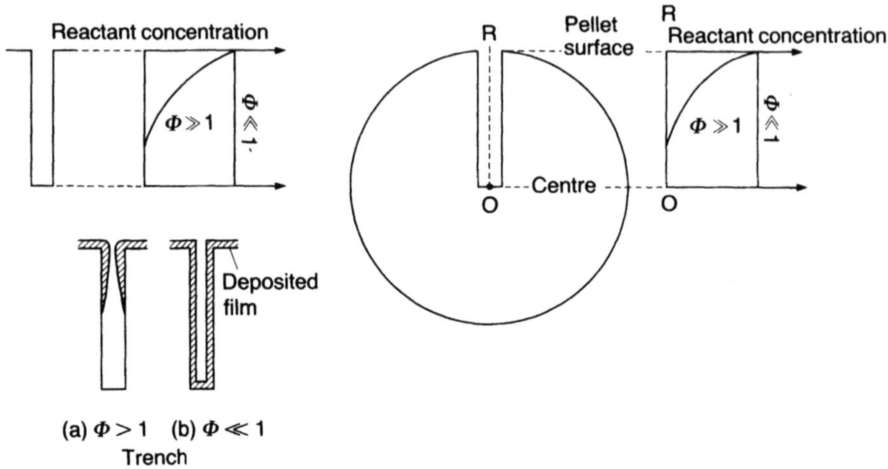

Figure 8.5 Relation between the conception of catalyst effectiveness and the concentration profile in a trench.

sensitive parameters which control absorption of oxygen into crystals. Where simulated results show unexpected eddies in the melt or instabilities in the interface temperature, new experiments can be undertaken to verify these effects. A detailed appraisal of the results of simulation studies is also useful in pointing out where improvements are possible in the model as well as suggesting how to improve the equipment.

It is very difficult to observe crystal growth since both the melt and solid are opaque. Further, *in situ* observation is not possible because of the risk of contamination by other metals and oxygen. The ability of being able to make predictions is a considerable advantage. Essentially, the purpose is to understand the phenomena so as to be able to improve operating conditions, provide guide-lines for deciding on operating conditions and to improve the processing equipment.

Thin film preparation (trench filling)

There is a tendency for the films to be thicker around the open mouth of a trench than at the bottom. In many ways, this is analogous to the concept of catalyst effectiveness factor or a reaction enhancement factor in absorption of gases [88, 89].

In a catalyst, as illustrated in Fig. 8.5, a reactant molecule diffuses into a porous heterogeneous catalyst and reacts to form a product molecule. This gives rise to a radial concentration profile of the reactant molecule in the catalyst pellet. The pore sizes found in a typical solid catalyst range from Å to microns, and so are smaller than, or about the same size as, the trench.

There is a straightforward, quantitative solution to this problem. For a first order irreversible reaction it depends on a dimensionless number, $m = L\sqrt{k/D}$ called the Thiele modulus. The concentration gradient is

$$\text{steep when } m \gg 1 \text{ and flat when } m \ll 1 \tag{8.3}$$

L is defined to be $2R$ (R = radius of equivalent sphere), D is the diffusion coefficient of the reactant in the catalyst and k is the first order intrinsic reaction rate constant. L and D can be measured or estimated, but the reaction rate constant k, has to be obtained by experiment. A practical method makes use of the modified modulus Φ ($= L^2r/DC_s$) where r is the reaction rate per unit catalyst volume, and C_s is the concentration of reactant at the catalyst surface. Both can be determined experimentally. The concentration profile will be:

$$\text{steep when } \Phi \gg 1 \text{ and flat when } \Phi \ll 1 \tag{8.4}$$

This analysis can be applied to the problem of filling the trench where L is the depth of the trench and D diffusion coefficient of the reactant. In this case, D is a function of the trench width and the mean free path length of the reactant molecule: r and C_s are the total deposition rate per unit volume of the trench and concentration of the reactant respectively. If Φ is greater than 1, the concentration profile has a large effect on the quality of the film. When Φ is less than 1 the quality is poor. The problem is how is this to be interpreted.

To produce a flat reactant concentration profile, some other factor needs to be included. One possibility is the presence of radicals derived from the gas phase which would have a higher probability of sticking to the surface. For deposition, C_s is the concentration of the radical rather than reactant and the value of Φ should therefore be greater than 1 [93, 94]. This shows that simulating the basic processes by means of a mathematical model makes it possible to interpret the underlying mechanism and so identify the circumstances under which poor quality film is formed. From this it is possible to find a way to improve the process.

The case study has been based on the DRAM, which is the commonest device. However, the indications are that the ratio of ad-hoc to commodity devices will tend to increase (in Japan). To avoid the huge investment costs of constructing large numbers of production lines requiring expensive instruments, an operational strategy needs to be established for manufacturing devices to various specifications, using a minimum number of production lines. A better understanding of the process mechanisms is indispensible in developing this kind of technology. Simulation based on a comprehensive model needs to be studied in relation to the deposition of thin films so as to be able to assess the performance in terms of the process operating conditions.

Revolutionary technology

There is considerable scope for applying chemical engineering principles to improve existing processes but there is even more potential for a breakthrough in VLSI technology. In trying to see what is involved, it is instructive to categorize the essential sub-processes which can be grouped under electronic circuits, structure of components, materials and manufacturing processes. There is a hierarchical relation between these since the electronic circuit (e.g. DRAM) governs the structure of the component (MOS) which in turn determines the material (SiO_2 thin films) and the manufacturing process (thermal oxidation).

Systems design depends on a complex set of inter-related factors which are best illustrated by reference to some examples. Consider a new material which is a high temperature super-conductor which requires a complete change in the sub-processes. A multilayered metallization process which reduces the access time to unit memory cells suggests a Josephson junction device. To create the necessary three-dimensional structures it is necessary to completely change the materials and processing. Any breakthrough would require the development of a new VLSI process which is significantly different from any presently used. It is the building of a total system for VLSI manufacture by combining all the stages which would represent the necessary innovative thinking.

Clearly, it is not possible to prejudge what might emerge from totally new applications of chemical engineering principles. There is certainly considerable potential for the application of systematic procedures for synthesis of the whole process which would draw together the principles of chemistry, physics and mechanics. The emphasis is on synthesis rather than analysis. Nevertheless, analysis does form part of the overall strategy since it is required in assessing the capability and potential of the synthesis step.

This argument can be illustrated by looking at aspects of the manufacturing process. The development of a new process is often combined with the simultaneous building of new equipment. For example, when the doping process was changed from diffusion to ion implantation, the instrumentation had to be completely replaced. The early work on this in the US was taken up in Japan and improved so that it could be operated on a much large scale. This pattern of development echoes what has happened in other Japanese industries. The consequence was that there has been an increased demand for chemical engineers because their skills make them well-suited to introducing radically new process technology. In conjunction with electronics, engineers, physicists and mechanical engineering specialists they have been able to make valuable contributions to VLSI technology.

Process simulation provides quantitative insight into the mechanisms which tend to limit the existing processes. For example, it has shown how to determine the required depth of a trench by reactive ion etching, how deep trenches can be cleaned by water and chemicals and how rapidly thin films can be deposited by chemical vapour deposition without forming a powder. It also makes it possible to carry out an appraisal of new advances in technology which will offer promise of being suitable for mass production.

Knowledge of the rate limiting step, provides an incentive to look for new ideas for dealing with the problem which might include new gases, equipment, materials, design principles or a fundamentally different approach to the manufacture process.

An example of this is the formation of a thin film. Equation (8.4) suggests a value of $L^2r/DC_s = 1$ can be used to establish whether there is a negligible concentration gradient of some key component during formation of the film. The depth of the trench is L but the diffusion coefficient, D, depends on the width of the trench as well. This means that it should be given as a device specification. The maximum growth rate of the film is then:

$$r_{\max} = \frac{DC_s}{L^2} \tag{8.5}$$

The increase in the film growth rate needs to be achieved without affecting the step coverage quality but is limited by $r_{\max}$. If this rate is incompatible with the processing time, a different approach will be required. This might be based on a reaction system where the growth rate is not dependent on the concentration of the reactant (the simulation assumes a first-order chemical reaction). Alternatively it might be possible to use selective deposition, as described in 8.1.4, where the film grows from the bottom of the trench and not on the sidewall of the trench. These features are not included in the simulation but might form the basis of new technology.

Table 8.1 attempts to show how present techniques, taken in conjunction with chemical engineering principles, can be used to create new processing methods.

8.2.2 Application of chemical engineering principles

Unlike laboratory scale operations, the complexities of large scale equipment make it almost impossible to describe what is happening in terms of a phenomenological models. One way of overcoming this is shown in Fig. 8.6. It is usually the case that large numbers of exploratory experiments

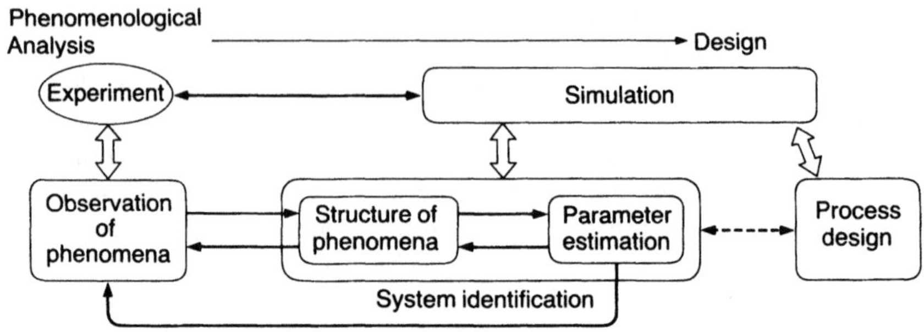

Figure 8.6 Chemical engineering approach to process design.

are needed to construct a model. Moreover the results can be influenced by the characteristics of the equipment. Modelling and simulation provides a means of overcoming these difficulties. Focusing on the chemical engineering aspects requires the final objective to be defined so that a strategy can be devised to reach it. At all stages, it is necessary to review constantly the procedures and update them when necessary. Thus, even if the concern is with research into basic material properties, it is necessary to give consideration to why they are required.

The approach which has proved successful in solving problems in chemical manufacturing can be extended to other industrial sectors and in particular the microscale operations in VLSI manufacture. However, there is a need to look at how operations can be scaled down to the molecular level where the units are mg and μm. The continuum world of classical physical-chemistry needs to be modified in order to analyse microscopic phenomena based on atoms and molecules. This requires a change in thinking from that which is appropriate to the petrochemicals industry so as to deal with the problems not only of VLSI technology but microbiology and fine powders.

Microscopic manipulation brings with it qualitative as well as quantitative problems. These arise in scale-up of fundamental research from the laboratory scale to production and in devising continuous processing procedures. Scale-up techniques for manufacturing processes are reasonably well established but are not appropriate for molecular processing. The VLSI processes considered in this book clearly fall into such a category (in common with microbiological systems), and a new approach to scale-up is therefore necessary. As noted previously, the changes taking place are still described by energy balance equations together with the rates of mass and heat transfer and reaction, whatever the scale of operation. The solution of these equations provides information on the concentration and temperature distribution and the fluid flow patterns. The fundamental approach therefore remains the same. It is the assumptions employed in modelling which need to be re-evaluated. An immediate difference is that it is no longer appropriate to regard diffusion, and thermal conduction as isotropic phenomena. Recognizing the intrinsic anistropy is crucial in crystal growth, since the heat and mass transfer processes in crystals have a very strong influence on the resulting material characteristics. Thus, in trying to grow single organic crystals having prescribed nonlinear optical properties, it is difficult to control the shape, because of the anisotropic growth rates arising from the difference surface energies in the crystal structure. Since the thermal conductivity also depends on the structural features, this also contributes to the anisotropic growth patterns.

To attack such problems it is necessary to look at the processes on a molecular scale. There is now considerable interest in adopting quantum mechanics to interpret rate data and statistical thermodynamic models to gain more insight. Unfortunately, while the physico-chemical process can be described, the equations describing the models have generally been impossible to solve up to now, although the ever-increasing computational power can be expected progressively to overcome many of the difficulties in due course.

The general nature of the problems can be appreciated by reference to the anisotropic single crystal growth referred to above. It is strongly influenced by the interaction between the solvent and components in the solution. In the case of m-chloronitrobenzene growing in acetone, the crystals are bounded by the (0, 1, 0), (1, 2, 0) and (1, 0, 1) crystallographic planes [94]. However, with a different solvent, the crystals take different shapes such as needles or powder. Essentially, it depends on the density of atoms on the surface. In addition, the specific atomic groups or radicals of organic compounds at the surface will give rise to different rates of crystal growth at the various crystal surfaces. The quantitative behaviour is not yet understood and obtaining reliable data will depend on good surface inspection measurement techniques. The task of the chemical engineer is to develop new procedures for the design of processing equipment for optimal operation, which will depend on employing microscopic analysis data in the design of production scale equipment.

The chemical engineering approach is well-suited to systems design because it has an interdisciplinary base. This makes it an effective tool in attacking problems in VLSI manufacture. In Japan, there has recently been a change in curriculum to address these kinds of problem which will be the basis of future technology. It will require the essential concepts, even if not the specific existing techniques, to be cast in a new context. The rest depends on the skills of the engineers using them.

This book has looked at VLSI process technology from the point of view of the chemical engineer. It has drawn attention to the need to understand the underlying structural relationships between the properties of materials and devices and the special considerations necessary to influence reactions at the surface of silicon wafers in order to be able to design processes and operate them efficiently. If this book proves to be useful in the tasks addressed, the authors will be very pleased.

Appendix A: Glossary

aligner: an apparatus that adjusts the position of a photo mask when a mask pattern is placed over a semiconductor wafer during lithography.

architecture: the characteristics and attributes of a computer. It may also include associated hardware and aspects of programming. In recent years it refers to the overall logic structure of the hardware system.

aspect ratio: the ratio of the material thickness to the pattern line width.

bipolar transistor: an integrated circuit that employs both active (transistors) and passive elements (resistances, capacitors) in one crystal substrate. The device contains two types of the carriers, positive holes and electrons.

Boolean algebra: the algebra formulated for mathematical logic and applied to logic, set theory, and particularly, the logic circuit design of computers.

captive maker: a maker producing semiconductors only for own use, for example, IBM and AT&T.

channelling: in ion implantation the deep penetration resulting from the beam entering parallel to the crystal planes and experiencing minimum interaction.

contact hole: holes cut in an oxide layer on the surface of a silicon substrate so as to make contact with the underlying gates, sources, drains, capacitors. It has an important effect on yield, characteristics, and reliability.

design rule: the standard for microfabrication in LSI elements, usually expressed in terms of line width. The design rule based on 1.3 μm corresponds to a 1 Mbit DRAM. This implicitly gives the sizes of each element and connections in the semiconductor cell.

dielectric breakdown phenomenon: the failure of a dielectric film by an electric discharge through the film due to a large increase in ionic electric charge on the surface of a dielectric thin film during the plasma ion etching process.

drain: the terminal that receives carriers from the source through the channel in a field effect transistor. The current carrier is controlled by the gate voltage.

electromigration: a phenomenon that carries away the atoms of aluminium by momentum exchange between thermally activated metal ions and conduction electrons. It may cause disconnection of a transistor.

electron cyclotron resonance: a phenomenon for efficiently absorbing microwaves. The frequency of the microwave coincides with the frequency of rotation of an electron in a magnetic field.

electron mobility: the drift velocity of free electrons in a semiconductor. The velocity of the electrons depends on the mobility and the strength of the electric field. It is 103 $cm^2/V\,s$ in silicon, and is several times that number in GaAs. It depends on the materials, impurity concentrations, electrons and positive holes.

epitaxial growth: a semiconductor manufacturing process in which a material is deposited from the gas phase and grows on the surface of a single crystal of the same material. The crystalline structure of the grown material exactly follows that of the original material. It is used in order to keep the impurity of a substrate at a suitable concentration in the working range.

fan out: the maximum number of logic gates that can be driven in parallel by the output of a single gate. When the branch number of a logic gate is greater than the number of the fan out, it is necessary to use the gate having the greatest driving power.

field oxidation: oxidation to form an insulator film which isolates each element. The thickest oxide film (800 nm thickness) is formed by steam oxidation, which also grows fastest.

forbidden band width: also called the energy gap, the width of an energy region between two allowed bands in which electrons cannot exist because of the interactions within the semiconductor.

gate array: an integrated circuit made by interconnecting the master wafers in the final stage of manufacture to provide the logic function requested by a customer. The master wafer is composed of a matrix of unconnected transistors and resistances on a single silicon chip.

gate breakdown: the destructive phenomenon of a gate function due to dielectric breakdown of the gate oxide film, which can happen during the reactive etching of polysilicon.

heterojunction: the junction between two kinds of semiconductors which differ in their doping level conductivities or alloy compositions, such as the junction between GaAs and GaAlAs.

hierarchical design: a method of building a system using stepwise design to go from a simple to a complex component. For instance, VLSI is designed in successively higher order levels: chip, block, cell, transistor.

impurity: a small proportion of foreign matter (e.g. indium) added to a pure semiconductor (e.g. silicon). Impurities, e.g. group III atoms, which accept electrons are called acceptors and cause the semiconductor (e.g. silicon) to be p-type. Impurities such as group V atoms provide free electrons called donors and cause the semiconductor (silicon) to be n-type. Thermal diffusion or ion implanation is used to insert the impurities.

intrinsic semiconductor: a semiconductor containing no impurities but in which some positive holes and electron pairs are created by thermal energy.

isolation: the electrical isolation of components in large scale integrated circuits to avoid interaction. They are isolated by p-n junctions or SiO_2.

Josephson device: a device using the Josephson effect which gives rise to current flow in a superconductor. Structures are classified as point contact bridge and tunnel connection type. A tunnel connection has an insulating film between a pair of superconductors and is used as a high frequency switching element by adjusting a magnetic field.

learning curve: an empirical rule based on experience which states that manpower and production costs decrease with experience. The resulting decreasing curve is called the learning curve. In the semiconductor manufacturing industry it is observed that if output doubles, production costs decrease by about 30%.

line balance: the balance between number of wafers manufactured and the production capacity of the equipment.

mask: a template used to print the circuit patterns on the surface of a silicon wafer. It is made by depositing thin opaque materials on a transparent supporting material.

parasitic element: an undesirable circuit element in a component or the interconnections, but not normally shown on circuit diagrams, e.g. stray capacitance/resistance or transistors.

passivation: stabilization of a semiconductor surface by deposition or growth of an inert layer so as to protect the surface from electrical, mechanical and chemical conditions in the environment.

pattern defect: a defect caused by particles sticking to a circuit pattern in the course of pattern formation.

pipeline processing: the processing method in which repeated calculations are divided into subunits which are partially processed in sequence.

polysilicon: a multicrystalline form of silicon. A thin silicon film grown by a CVD process in VLSI manufacturing or silicon rods produced by the Siemens method for single crystal silicon manufacture.

positive hole: an electron deficiency in the valence structure of a semiconductor caused by implanting impurity atoms having smaller valence than the surrounding atoms of the semiconductor. It is capable of 'moving' through the crystal and forms an effective current carrier with a positive charge.

reactive ion etching (RIE): an etching process using physical attack by ion particles and chemical reaction of ion radicals from a gas plasma. It is an example of a dry etching processes.

Schottky barrier: an electronic potential barrier occurring in a metal–semiconductor interface, where the metal contacts the semiconductor. This phenomenon was theoretically explained by Schottky in 1938. When a metal is in contact with an n-type semiconductor, if the semiconductor is negatively charged, an electric current flows through the interface; if the metal is negatively charged, there is no current.

secondary electron: an electron emitted from a material as a result of being struck by other high velocity electrons. It has low energy (<50 ev) and the amount depends on the surface because it occurs only near the surface (<10 nm).

selective diffusion: an additional method of introducing impurities by a solid-state diffusion technique into desired regions of a semiconductor wafer, depending on the selected concentration and pattern.

selective (etching) ratio: the ratio of the reaction rate of etched material to that of the resist materials and substrates which are not etched.

self align: a film formed on a device in place of a mask. It eliminates the processing step of mask adjustment, and so increases the accuracy of fabrication.

semiconductor component per equipment rate: the ratio of the number of semiconductor components used in a piece of equipment to the number of such units produced.

Shannon, Claude Edward: an electrical engineer at the Bell Telephone Laboratory in the US who was a pioneer in developing information theory.

sheet resistance: the electrical resistance of a diffuse layer. It is important to evaluate the diffusion depth and impurity concentration profile.

silicon cycle: the production cycle in the semiconductor industry. The silicon cycle has a period of approximately 3–4 years.

single wafer method: a method of treating silicon wafers one by one in the fabrication process. It is used in lithography, etching, and ion implantation. In batch processing 25 wafers constitute one lot.

source: the terminal which supplies carriers to the drain channel in a field effect transistor, i.e. the source of holes (p channel) flowing in the channel.

sputtering: a thin film formation technique in which molecules are ejected from the surface of the bulk source on a cathode by ion bombardment (normally Ar^+) in a glow discharge plasma. It is then deposited on the substrate of an anode as a thin film.

stacking fault: a defect in crystal structure resulting from incorrect stacking of the close packed atom planes. If vacancies are produced in these planes, it is a fault type; if atoms are inserted into the interstitial sites, it is a fault of the insertion type.

standard cell: cells which are small or medium scale integrated circuits standardized as unit function cells for common use in designing logic circuits. Semicustom integrated circuits are manufactured by combining such cells.

threshold voltage: the input voltage where a p-n junction begins to conduct current in a semiconductor circuit or device.

through hole: a hole for interlayer connection in multi-level assemblies.

trench structure: a three-dimensional structure in which the electrical capacitance is increased by using the side of a trench formed during micro fabrication.

tunnel effect: the passages of electrons through a (sufficiently thin) insulating barrier: it is a quantum mechanical phenomenon. In a p-n junction, electrons pass in either direction between the conduction band in the n region and the valence band in the p region, if both resistances are small enough.

undercut: an etching below a resist material. Although it offers protection it gives rise to fabrication errors.

Von Neumamn, John: American mathematician who developed the theoretical principles of serial computation.

Appendix B: List of Abbreviations

ASCP (application specific custom product): special integrated circuits developed to meet the specifications for a particular application of the customer. Also called user specific integrated circuit (USIC).

ASIC (application specific integrated circuit): specific integrated circuits used only for a particular application.

ASSP (application specific standard product): standard integrated circuits used for a special application by multiple customers e.g. ASCP.

Bi-MOS (bipolar metal oxide semiconductor): an integrated circuit employing both analogue and digital elements in one crystal substrate.

CIMS (computer integrated manufacturing system): a system which coordinates design, development, production, process control, purchasing, distribution, cost accounting, marketing and support functions in a manufacturing enterprise. This system is a more advanced computer integrated system than FMS.

CMOS (complementary metal oxide semiconductor): a MOS that employs both n-channel and p-channel MOS transistors in one crystal substrate. The gates and drains of the transistors are connected together.

CVD (chemical vapour deposition): an thin film formation process in which a single crystal or an insulating film grows on the substrate by chemical reaction with gas molecules from a high temperature gas.

CZ method (Czochralski method): a single crystal growth technique in which a rotating seed crystal is dipped into a melt container where a polycrystalline material is melted and then slowly withdrawn.

DRAM (dynamic random access memory): a RAM in which memory contents disappear within a short time period if the electric power source is cut off.

EEPROM (electrically erasable programmable read only memory): a non-volatile ROM in which the memory contents can be electrically written and rewritten by users. The insulating film between the floating gate and drain is 10–20 nm, and electrons can transfer by tunnelling.

EPROM (erasable programmable read only memory): a non-volatile ROM in which programs can be written by a user. It holds a thin insulating film between floating and control gate, and is written on to by putting hot electrons into the floating gate by applying a high voltage between the control and a drain. The contents are cleared by subjecting it to UV light or an electric discharge of the floating gate.

FET (field effect transistor): a transistor in which the output current is controlled, amplified, and switched by the effect of an electrical field derived from an input voltage. The device contains either positive holes or electrons. Also called a monopolar transistor.

FMS (flexible manufacturing system): a manufacturing system in which manufacturing functions can be changed quickly according to the needs of the market. In this system one computer controls many NC machines, transfer lines, automatic storage and retrieval systems.

FZ method (floating zone method): a crystal growing method in which a polycrystalline rod is kept in a vertical position, melted by high frequency wave, cooled and converted into a single crystal. The molten region is supported by surface tension or electromagnetic levitation.

LOCOS (local oxidation of silicon): a selective oxidation method using a silicon nitride film as an oxidation mask.

MOS (metal oxide semiconductor): a three-layer device attached to an insulating oxide film between a semiconductor substrate and a metal gate electrode.

MRI (magnetic resonance imager): computer tomography based on nuclear magnetic resonance, also called NMR CT. It generates a tomogram in any direction and there are no ill-effects of the radiation, unlike in X-ray tomography.

NIES: newly industrialized economics.

n-MOS (n-channel metal oxide semiconductor): a MOS in which a channel is formed by electrons moving between source and sink.

p-CVD (plasma chemical vapour deposition): a CVD process in which chemical ions in a gas plasma are used to create an electric discharge under reduced pressure.

p-MOS (p-channel metal oxide semiconductor): a MOS in which a channel is formed by positive holes moving between source and sink.

PROM (programmable read only memory): a non-volatile ROM in which a program can be written only once.

PSG (phospho silicon glass): silicate glass containing phosphate used as an interlayer insulator. It is used to form thin films on the side wall of etchings because it flows more easily than polysilicon.

RAM (random access memory): a volatile storage device in which information can be written into and read from an arbitrary location in any order and the access time is independent of the location of the information.

RIE (reactive ion etching): see Glossary.

ROM (read only memory): a read-only non-volatile storage device in which the stored information does not disappear if the electric power source is cut off.

SOI (silicon on insulator): a process in which single crystal silicon grows on an insulating film.

SRAM (static random access memory): a volatile RAM in which memory contents do not disappear unless the electric power source is cut off.

WSTS (world semiconductor trade statistics): the voluntary statistics agency of the world's semiconductor makers. About sixty makers from Japan, USA, and Europe belong to the agency. It collects monthly statistics on acceptance of orders and shipments and makes and analysis of mid-term demand.

References

[1] Bardeen, J., and Brattain, W. H. (1948) *Phys. Rev.*, **78**, 230.
[2] Tsuji, S., Matsumura, H., and Ohuchi, H. (1986) *Hitachi Rev.*, **35**, 207.
[3] Moore, G. E. (1975) *Proc. Kodak Microelec. Seminar, 5*. This figure has been revised to take note of recent developments.
[4] Sugawara, K. (1987) *Kagaku Kogaku (Chemical Engineering of Japan)*, **51**, 336.
[5] Hososaka, K., and Anzai, A. (1982) *Hitachi Hyoron*, **64**, 525.
[6] Owada, N., *et al.* (1985) *Proc. 2nd Int. IEEE VISI Multilevel Interconnection Conf.*, pp. 173.
[7] Kimbara, K. (1986) *2nd Annual Int. Trade Partners Conf.*
[8] Matsuo, H., Morioka, T., Saito, N., and Ishikawa, K. (1990) *Hitachi Hyoron*, **72**, 219.
[9] Einspruch, N. G. (ed.) (1981–1987) *VISI Electronics Microstructure Science*, **1–15**, Academic Press.
[10] Millman, Jacob, and Grabel, Arvin (1987) *Microelectronics*, 2nd edn, McGraw Hill.
[11] Korn, G. A. (1977) *Microprocessors and small digital computer systems for engineers and scientsts*, McGraw Hill.
[12] Einspruch, N. G. (ed.) (1985) *VLSI handbook*, Academic Press.
[13] Barb, D. F. (Ed.) (1980) *Very large scale integration (VLSI)-fundamental and applications – Springer-Verlag.*
[14] Barna, A., Porat, D. I. (1979) *Introduction to microcomputers and microprocessors*, John Wiley & Sons.
[15] Shinagawa, Yoshiya (1972) *Brain and computer* (in Japanese), Chuoukoron Sha.
[16] Kayama, Susumu (1985) (ed. T. Sugano), *Super-high speed MOS devices* (in Japanese), Baifuukan.
[17] Nagata, Minoru (1985) (ed. T. Sugano), *Super-high speed bipolar device* (Japanese), Baifuukan.
[18] Nakagawa, K., *et al.* (1989) *Appl. Phys. Lett.*, **54**(19).
[19] Akasaka, Y. (1986) *Proc. IEEE*, **74**.
[20] Sekigawa, T., *et al.* (1985) *Solid-state Electronics*, **27**, 827.
[21] Kung, H. T. (1974) *Proc. Caltech Conf. VLSI.*, 65–91.
[22] Suzuki, M., *et al.* (1987) *IEEE J. Solid-state Circuits*, **24**(5), 1233.
[23] Tanaka, H., *et al.* (1987) *ISSC*.
[24] Seki, S., *et al.* (1988) *IEDM Tech. Digest*, 770.
[25] Matsumoto, K., *et al.* (1984) *IEEE Electron Device Lett.*, 20.
[26] Su, S. L., *et al.* (1983) *J. Appl. Phys.*, **54**(11), 6725.

[27] Yokoyama, N., *et al.* (1984) *Technical Digest of IEDM 84*, 532.
[28] Yokoyama, N., *et al.* (1985) *J. Jap. Appl. Phys.*, **24**(11), 853.
[29] Ishibashi, T., *et al.* (1988) *IEEE Trans. Electron Devices*, **35**, 401.
[30] Hatano, Y., *et al.* (1990) *IEEE M*, 40.
[31] Nishino, T., *et al.* (1989) *IEEE Electron Device Lett.*, **10**, 61.
[32] Harada Y., *et al.* (1987) *Proc. IEEE 77*, 1280.
[33] Streetman, B. G. (1990) *Solid state electric devices*, 3rd edn, Prentice Hall, Englewood Cliffs, New Jersey 07632, Sec. 9.
[34] Hokari, Y. (1986) *ULSI*, **2**(8), 14.
[35] May, T. C., and Woods, M. H. (1978) *Proc. 1978 Reliability Physics Symp.*, 33.
[36] Deal, B. E., and Grove, A. S. (1965) *J. Appl. Phys.*, **36**, 3770.
[37] Cabrera, N., and Mott, N. F. (1948) *Rept. Progr., Phys.*, **12**, 163.
[38] Osburm, C. M., and Ormond, D. W. (1992) *J. Electrochem. Soc.*, **119**, 597.
[39] Yamabe, K., and Taniguchi, K. (1985) *IEEE Trans. E. D.*, **32**, 423.
[40] Ziegler, Z. J., Biersack, J. P., and Littmark, V. (1983) *Proc. of Int. Ion Engng Cong. ISIAT' 83*, 1861.
[41] Mayer, J. W., Eriksson, L., and Davies, J. A. (1970) *Ion implantation in semiconductors; Si and Ge*, Academic.
[42] Lindhard, J., Scharff, M., and Schiott, H. E. (1963) *Mat. Frys. Medd. Dan. Vid. Selsk.*, **33**, 14.
[43] Gibbons, J. F., and Myhrorie, S. (1973) *Appl. Phys. Letters.*, **22**, 568.
[44] Hofker, W. H. (1975) *Phillips Res. Rep.*, **8**, 41.
[45] Winterborn, T. K. B., Sigmond, P., and Sanders, J. B. (1970) *Mat, Fys. Medd. Dan. Vid. Selsk*, **37**, 1.
[46] Ryssel, H., and Hoffmann, K. (1983) *Process and device simulation for MOS–VLSI Circuits*, Hague, Martinus Nijihoff.
[47] Furukawa, S., Matsumura, H., and Ishiwara, H. (1972) *Jpn. J. Appl. Phys.*, **11**, 134.
[48] Hetcher, J., Nrayan, J., and Holland, O. W. (1981) *Inst. Phys, Conf. Ser.*, **60**, 295.
[49] Williams J. S., and Short, K. T. (1982) *Metastable phases by ion implantation*, North-Holland.
[50] Breiland, W. G., Ho, P., and Coltrin, M. E. (1986) *J. Appl. Phys.*, **60**, 1505.
[51] Vossen, J. L., and Kern, W. (1978) *Thin film processes*, Academic Press.
[52] Tanaka, K. (1989) *Glow-discharge hydrogenated amorphous silicon*, KTK Scientific Publishers, Tokyo.
[53] Chapman, B. (1980) *Glow-discharge processes*, John Wiley & Sons.
[54] For example, Levenspiel, O. (1972) *Chemical Reaction Engineering* (final edn.) John Wiley & Sons.
[55] Hirschfelder, Curtis, Bird, (1954) *Molecular theory of Gases and liquid*, John Wiley & Sons.
[56] Kinjo, N., Ogata, M., Nishi, K., and Kaneda, A. (1989) *Advances in Polymer Science*, **88**, 1, Springer-Verlag Berlin Heidelberg.
[57] Kinbara, K. (1988) *Nihon Handotai-Seizosochi Kyokai Kaishi*, **9**, 24.
[58] Murakami, G., Nishi, K., Kawai, S., Ogata, S., and Kaneda, A. (1987) *Hitachi Hyoron*, **69**, 609.
[59] Reimer, L. (1985) *Scanning Electron Microscopy*, Springer-Verlag, Berlin.
[60] Nyyssonen, D. (1977) Linewidth measurement with an optical microscope: The effect of operating on the image profile, *Appl. Opt.*, **16**, 2223.
[61] Hamashima, M., Kato, K., and Ishizeki, T. (1988) Optical measurement of half micron critical dimensions, *Proc. SPIE 921*, 92.

[62] Miyoshi, M., and Yamazaki, Y. (1986) Topographic contrast in the linewidth measurement with a scanning electron microscope, *J. Electron Microscopy*, **35**(2), 118; Miyoshi, M., Kanoh, M., Yamaji, H., Okumura, K. (1986) A precise and automatic very large scale integrated circuit linewidth measurement method using a scanning electron microscope, *J. Vac. Sci. Technol.*, **B4**(2), 493.
[63] Miyoshi, M., Kanoh, M., Yamaji, H., and Okamura K. (1986) A precise and automatic VLSI Circuit linewidth measurement method using SEM, *J. Vac. Sci. Technol.*, **B4**(2), 493.
[64] Born, M., and Wolf, E. (1964) *Principle of Optics*, Pergamon Press, Oxford.
[65] Tolliver, D. (1984) Contamination control: new dimensions in VLSI manufacturing, *Solid State Technology*, **27**(3), 129.
[66] Chopra, K. (1979) *Thin film phenomena*, Robert E. Kriger Publishing, Huntington.
[67] Benninghoven, A., Giber, J., Laszlo, J., Riedel, M., and Werner, H. (1982) Secondary ion mass spectrometry; SIMS III, Springer-Verlag, Berlin.
[68] Wittmack, K. (1982) Design and performance of quadruple-based SIMS instruments: a critical review, *Vacuum*, **32**(2), 65.
[69] Ito, Y. (1985) *Moru*, **23**(7), 64 (in Japanese).
[70] The Chemical Society of Japan (1986) *Kagaku Binran (Applied Chemistry, I, II)*, Maruzen, Tokyo (in Japanese).
[71] Tokuyama, I., and Hashimoto, K. (1985) *MOS LSI production technology*, Nikkei-McGraw-Hill Co., Tokyo (in Japanese).
[72] Harada, M. (1985) *Semiconductor World*, **4**, 122 (in Japanese).
[73] Kato, M. (1988) *Moru*, **26**(4), 28 (in Japanese).
[74] (1988) *Gasu Rebyu*, **160** (in Japanese).
[75] Taito (1986) *Moru*, **24**(3), 27 (in Japanese).
[76] Koike, K. (1986) *Moru*, **24**(2), 57 (in Japanese).
[77] Suzuki, M., *et al.* (1986) *The Hitachi Hyoron*, **68**, 737.
[78] Sinoda, M., *et al.* (1986) *Semiconductor World*, **6**, 69.
[79] Parikh, M., *et al.* (1984) *Solid State Technology*, **27**, 50.
[80] Bantaz, G., *et al.* (1986) *Ibid.*, **29**, 54.
[81] Takahashi, T. (1986) *ULSI*, **2**(7), 56.
[82] WHO Environmental Health Criteria 50 (1985).
[83] WHO Environmental Health Criteria 31 (1984).
[84] Tsubouchi, H. *et al.* (1977) *Japan J. Appl. Phys.*, **16**, 855.
[85] Champagne, R., and Toole, M. (1977) *Solid State Technology*, **20**, 61.
[86] Zeto, R. J., Korolkoff, N. O., and Marhsall, S. (1979) *Solid State Technology*, **22**, 62.
[87] Gourrier, S., and Bacal, M. (1981) *Plasma chemistry and plasma processing*, **1**, 217.
[88] Satterfield, C. N. (1970) *Mass transfer in heterogeneous catalyst*, MIT, Mass.
[89] Froment, G. F., and Bischoff, K. B. (1990) *Chemical Reactor Analysis and Design*, John Wiley & Sons, Ch. 6.
[90] Joyce, B. D. and Baldray J. A. (1962) *Nature*, **195**, 485.
[91] Shaw, J. M., Amic, J. A. (1971) *Solid State Tech.*, **14**, 53.
[92] Watanabe, K., and Komiyama, H. (1990) *J. Elect. Chem. Soc.*, **137**, 1222.
[93] Sawada, Okazawa, Tanaka, and Tsukijima: (1986) *Proc. 25th Heat Transfer Symp.*, Japan, B211, June.
[94] Matsuoka, M., and Ueda R. (1986) *Zairyo Gijutsu*, **4**, 38.

[95] Antoniadis, D. A. *et al.* (1979) *IEEE J. Solid-state circuits*, **14**, 412.
[96] Dutton, R. W. *et al.* (1983) *IEEE Trans Electron Dev.*, **30**, 968.
[97] Oldham, W. G., *et al.* (1979) *IEEE Trans Electron Dev.*, **26**, 717.
[98] Ongo, M., *et al.* (1985) *Tech. Res. Rept. Institute of Gectonics, Information and Communications Engineers*, Japan, SSD 85–64 (Sept); Ongo, M., *et al.* (1987) *IEEE Trans C. A. D. of ICAS*, **6**, 439.
[99] Toyabe, T., *et al.* (1978) *IEEE Trans. Electon Dev.*, **25**, 825.
[100] Toyabe, T., *et al.* (1988) *IEEE Trans. Electron Dev.*, **32**, 2038.
[101] Greenfield, J. A., *et al.* (1980) *IEEE Trans. Electron Dev.*, **27**, 1520.
[102] Chamber lain, S. G. (1981) *IEDM Tech. Dig.*, 592.
[103] Buturla, E. M. (1981) *IBM J. Res. Develop.*, **25**, 218.
[104] Jeda (1984) *Trans Inst. Electonics, Information and Communication Engineers*, Japan. **67**, 825.
[105] Engl, W. L., *et al.* (1982) *IEEE Trans C. A. D. of ICAS* **1**, 85.
[106] Nagel, L. W. (1973) *19m Midloost Symp. Circiut Theory*, April.
[107] Kalter, H. and Van de Ven, E. P. G. T. (1978/9) *Philips Tech. Rev.* **35**(7/8).
[108] Hiratsuka, Y. (1987) *Know-how of Precise Cleaning and Measures against Particle Contamination*, paper presented at Realize Company Seminar, p. 128 (in Japanese).
[109] Maeda, K. (1983) *New LSI Process Technology*, Kogyo Chosakai, Tokyo (in Japanese).
[110] Kern, W. and Puotinen, D. A. (1970) *RCA Review*, June, p. 187.
[111] *Annual Book of ASTM Standards* (1980) Part 43, F612–80.
[112] Sze, S. M. (1981) *Physics of Semiconductor Devices*, 2nd edn, John Wiley & Sons, p. 184.

Index

Entries in *italics* refer to tables; entries in **bold** refer to figures.

GPSR Compliance

The European Union's (EU) General Product Safety Regulation (GPSR) is a set of rules that requires consumer products to be safe and our obligations to ensure this.

If you have any concerns about our products, you can contact us on ProductSafety@springernature.com

In case Publisher is established outside the EU, the EU authorized representative is:

Springer Nature Customer Service Center GmbH
Europaplatz 3
69115 Heidelberg, Germany

Batch number: 10607222

Printed by Printforce, the Netherlands